MATH WORKBOOK

Emma.School

Dear buyer,

I want to take a moment to thank you for your recent purchase of my math workbook. Your support means the world to me, and I am so grateful for your decision to invest in your or your kids math education.

If you have any questions or feedback about the workbook, please don't hesitate to reach out to me this is my email: contact@emmaworksheets.com. I am always looking for ways to improve my materials and make them as useful as possible for my buyers.

You can visit my author page " Emma.School " on Amazon if you need any other workbook, there are more than 100 workbooks that can help you or your children improve their skills in mathematics

and If you are interested in free worksheets, you can visit my website:

www.emmaworksheets.com

My hope is that my workbook will be a valuable tool for you as you learn and practice essential math skills.

Once again, thank you for your purchase and for your commitment to improving your math skills. I wish you all the best in your studies and hope that you find the workbook to be a valuable resource.

Sincerely,

ADDITION EXERCISES

1) 26 + 57 = ...	2) 63 + 99 = ...	3) 50 + 45 = ...	4) 65 + 73 = ...	5) 15 + 62 = ...
6) 37 + 41 = ...	7) 49 + 11 = ...	8) 88 + 28 = ...	9) 18 + 47 = ...	10) 21 + 49 = ...
11) 12 + 21 = ...	12) 55 + 79 = ...	13) 79 + 21 = ...	14) 95 + 46 = ...	15) 97 + 68 = ...
16) 32 + 65 = ...	17) 94 + 12 = ...	18) 26 + 16 = ...	19) 58 + 74 = ...	20) 42 + 39 = ...
21) 72 + 96 = ...	22) 88 + 98 = ...	23) 12 + 77 = ...	24) 85 + 68 = ...	25) 91 + 90 = ...
26) 48 + 17 = ...	27) 44 + 78 = ...	28) 70 + 14 = ...	29) 12 + 83 = ...	30) 29 + 39 = ...
31) 49 + 29 = ...	32) 75 + 46 = ...	33) 40 + 71 = ...	34) 45 + 47 = ...	35) 84 + 82 = ...

1) 14 + 11	2) 63 + 11	3) 66 + 81	4) 55 + 15	5) 27 + 14
6) 21 + 74	7) 68 + 25	8) 25 + 50	9) 29 + 16	10) 73 + 78
11) 94 + 88	12) 12 + 90	13) 35 + 83	14) 80 + 87	15) 48 + 70
16) 85 + 53	17) 45 + 86	18) 27 + 23	19) 74 + 40	20) 58 + 69
21) 38 + 98	22) 78 + 77	23) 62 + 37	24) 53 + 89	25) 22 + 42
26) 20 + 39	27) 35 + 26	28) 98 + 72	29) 76 + 82	30) 11 + 66
31) 66 + 56	32) 28 + 67	33) 39 + 82	34) 97 + 12	35) 81 + 15

1) 75 + 19 = ...	2) 99 + 65 = ...	3) 94 + 61 = ...	4) 29 + 61 = ...	5) 57 + 32 = ...
6) 19 + 98 = ...	7) 74 + 33 = ...	8) 34 + 31 = ...	9) 96 + 87 = ...	10) 85 + 66 = ...
11) 51 + 15 = ...	12) 39 + 33 = ...	13) 78 + 33 = ...	14) 99 + 19 = ...	15) 51 + 24 = ...
16) 93 + 50 = ...	17) 49 + 53 = ...	18) 67 + 96 = ...	19) 60 + 45 = ...	20) 29 + 44 = ...
21) 29 + 39 = ...	22) 81 + 69 = ...	23) 93 + 82 = ...	24) 16 + 87 = ...	25) 24 + 58 = ...
26) 41 + 63 = ...	27) 32 + 51 = ...	28) 58 + 71 = ...	29) 78 + 76 = ...	30) 86 + 71 = ...
31) 61 + 20 = ...	32) 21 + 43 = ...	33) 37 + 68 = ...	34) 59 + 30 = ...	35) 65 + 37 = ...

1) 79 + 94	2) 79 + 13	3) 47 + 48	4) 19 + 47	5) 90 + 55
6) 49 + 24	7) 28 + 60	8) 96 + 50	9) 60 + 36	10) 69 + 19
11) 86 + 13	12) 21 + 69	13) 39 + 93	14) 84 + 38	15) 15 + 72
16) 41 + 89	17) 19 + 16	18) 13 + 45	19) 48 + 94	20) 57 + 57
21) 67 + 12	22) 43 + 93	23) 10 + 87	24) 44 + 83	25) 72 + 59
26) 62 + 77	27) 11 + 76	28) 83 + 21	29) 74 + 53	30) 63 + 80
31) 42 + 28	32) 23 + 59	33) 16 + 22	34) 57 + 61	35) 12 + 65

1) 142 + 90	2) 149 + 143	3) 35 + 50	4) 139 + 62	5) 23 + 61
6) 185 + 22	7) 54 + 170	8) 109 + 60	9) 71 + 185	10) 46 + 189
11) 22 + 193	12) 197 + 35	13) 71 + 152	14) 166 + 172	15) 98 + 60
16) 150 + 175	17) 135 + 60	18) 68 + 70	19) 141 + 32	20) 23 + 88
21) 34 + 107	22) 26 + 30	23) 114 + 88	24) 158 + 48	25) 165 + 146
26) 44 + 152	27) 157 + 127	28) 198 + 150	29) 93 + 124	30) 83 + 84
31) 84 + 167	32) 164 + 163	33) 111 + 20	34) 87 + 194	35) 112 + 49

1) 195 + 180 = ...	2) 199 + 162 = ...	3) 93 + 173 = ...	4) 117 + 13 = ...	5) 100 + 90 = ...
6) 156 + 160 = ...	7) 22 + 147 = ...	8) 113 + 193 = ...	9) 80 + 142 = ...	10) 194 + 72 = ...
11) 102 + 98 = ...	12) 42 + 154 = ...	13) 109 + 172 = ...	14) 124 + 67 = ...	15) 27 + 183 = ...
16) 40 + 193 = ...	17) 92 + 124 = ...	18) 31 + 29 = ...	19) 28 + 55 = ...	20) 169 + 145 = ...
21) 70 + 24 = ...	22) 60 + 60 = ...	23) 57 + 106 = ...	24) 130 + 116 = ...	25) 17 + 111 = ...
26) 130 + 31 = ...	27) 141 + 64 = ...	28) 136 + 107 = ...	29) 147 + 55 = ...	30) 64 + 120 = ...
31) 186 + 117 = ...	32) 60 + 163 = ...	33) 79 + 98 = ...	34) 126 + 94 = ...	35) 175 + 35 = ...

1) 151 + 215 = ...	2) 279 + 132 = ...	3) 252 + 266 = ...	4) 305 + 200 = ...	5) 184 + 139 = ...
6) 413 + 283 = ...	7) 383 + 139 = ...	8) 278 + 295 = ...	9) 390 + 273 = ...	10) 262 + 171 = ...
11) 253 + 162 = ...	12) 482 + 161 = ...	13) 237 + 265 = ...	14) 402 + 299 = ...	15) 349 + 254 = ...
16) 500 + 216 = ...	17) 215 + 204 = ...	18) 382 + 203 = ...	19) 355 + 275 = ...	20) 242 + 266 = ...
21) 138 + 140 = ...	22) 444 + 138 = ...	23) 310 + 149 = ...	24) 430 + 250 = ...	25) 411 + 283 = ...
26) 421 + 213 = ...	27) 192 + 107 = ...	28) 180 + 264 = ...	29) 348 + 128 = ...	30) 426 + 211 = ...
31) 131 + 203 = ...	32) 452 + 188 = ...	33) 178 + 140 = ...	34) 351 + 137 = ...	35) 295 + 286 = ...

1) 482 + 111	2) 363 + 255	3) 106 + 174	4) 443 + 173	5) 387 + 210
6) 130 + 159	7) 169 + 195	8) 404 + 271	9) 251 + 273	10) 339 + 106
11) 436 + 258	12) 383 + 263	13) 372 + 273	14) 112 + 253	15) 283 + 197
16) 244 + 248	17) 249 + 126	18) 105 + 176	19) 377 + 195	20) 223 + 204
21) 115 + 127	22) 238 + 150	23) 373 + 157	24) 159 + 289	25) 195 + 194
26) 196 + 255	27) 367 + 263	28) 395 + 133	29) 239 + 268	30) 138 + 130
31) 236 + 119	32) 462 + 147	33) 176 + 167	34) 469 + 108	35) 284 + 177

1) 235 + 144 = ...	2) 335 + 264 = ...	3) 112 + 245 = ...	4) 457 + 241 = ...	5) 380 + 187 = ...
6) 202 + 214 = ...	7) 467 + 298 = ...	8) 342 + 153 = ...	9) 246 + 230 = ...	10) 346 + 171 = ...
11) 225 + 205 = ...	12) 259 + 280 = ...	13) 120 + 196 = ...	14) 233 + 179 = ...	15) 416 + 106 = ...
16) 344 + 220 = ...	17) 368 + 236 = ...	18) 463 + 173 = ...	19) 293 + 247 = ...	20) 244 + 124 = ...
21) 155 + 129 = ...	22) 371 + 245 = ...	23) 470 + 175 = ...	24) 224 + 125 = ...	25) 282 + 116 = ...
26) 499 + 191 = ...	27) 448 + 200 = ...	28) 499 + 263 = ...	29) 428 + 238 = ...	30) 408 + 210 = ...
31) 257 + 199 = ...	32) 467 + 234 = ...	33) 429 + 232 = ...	34) 459 + 148 = ...	35) 133 + 148 = ...

1)	2)	3)	4)	5)
247 + 133 = ...	250 + 157 = ...	431 + 294 = ...	159 + 243 = ...	138 + 175 = ...

6)	7)	8)	9)	10)
173 + 210 = ...	183 + 227 = ...	367 + 286 = ...	392 + 211 = ...	496 + 238 = ...

11)	12)	13)	14)	15)
496 + 233 = ...	291 + 253 = ...	330 + 288 = ...	434 + 254 = ...	383 + 140 = ...

16)	17)	18)	19)	20)
141 + 134 = ...	390 + 183 = ...	367 + 298 = ...	241 + 277 = ...	442 + 125 = ...

21)	22)	23)	24)	25)
260 + 115 = ...	448 + 187 = ...	319 + 118 = ...	107 + 175 = ...	194 + 167 = ...

26)	27)	28)	29)	30)
399 + 203 = ...	120 + 205 = ...	222 + 240 = ...	375 + 222 = ...	385 + 277 = ...

31)	32)	33)	34)	35)
153 + 175 = ...	103 + 279 = ...	210 + 103 = ...	431 + 259 = ...	486 + 236 = ...

1) 27 oranges are in the basket. Two more oranges are put in the basket. How many oranges are in the basket now?

....

2) David has 29 pears and Billy has 11 pears. How many pears do David and Billy have together?

....

3) Some apples were in the basket. 46 more apples were added to the basket. Now there are 72 apples. How many apples were in the basket before more apples were added?

....

4) Five red plums and 34 green plums are in the basket. How many plums are in the basket?

....

5) 62 mangoes were in the basket. 28 are red and the rest are green. How many mangoes are green?

....

6) 38 bananas were in the basket. More bananas were added to the basket. Now there are 56 bananas. How many bananas were added to the basket?

....

7) Jennifer has 14 more peaches than Marin. Marin has 39 peaches. How many peaches does Jennifer have?

....

8) Some plums were in the basket. 39 more plums were added to the basket. Now there are 76 plums. How many plums were in the basket before more plums were added?

....

1) 42 red plums and 49 green plums are in the basket. How many plums are in the basket?

....

2) 67 mangoes were in the basket. 29 are red and the rest are green. How many mangoes are green?

....

3) 46 apples were in the basket. More apples were added to the basket. Now there are 48 apples. How many apples were added to the basket?

....

4) Marcie has 27 more bananas than Ellen. Ellen has 35 bananas. How many bananas does Marcie have?

....

5) Some pears were in the basket. 38 more pears were added to the basket. Now there are 82 pears. How many pears were in the basket before more pears were added?

....

6) Billy has 38 peaches and Donald has 30 peaches. How many peaches do Billy and Donald have together?

....

7) 18 oranges are in the basket. Nine more oranges are put in the basket. How many oranges are in the basket now?

....

8) 70 plums were in the basket. 20 are red and the rest are green. How many plums are green?

....

1) Donald has 23 peaches and Allan has 46 peaches. How many peaches do Donald and Allan have together?

....

2) 49 apples are in the basket. 42 more apples are put in the basket. How many apples are in the basket now?

....

3) 18 bananas were in the basket. More bananas were added to the basket. Now there are 22 bananas. How many bananas were added to the basket?

....

4) Some mangoes were in the basket. 46 more mangoes were added to the basket. Now there are 79 mangoes. How many mangoes were in the basket before more mangoes were added?

....

5) Ellen has 32 more oranges than Sharon. Sharon has 23 oranges. How many oranges does Ellen have?

....

6) Nine red pears and 23 green pears are in the basket. How many pears are in the basket?

....

7) 18 plums were in the basket. 11 are red and the rest are green. How many plums are green?

....

8) 41 red plums and 34 green plums are in the basket. How many plums are in the basket?

....

1) 91 mangoes were in the basket. 41 are red and the rest are green. How many mangoes are green?

....

2) Marin has 24 more peaches than Jackie. Jackie has nine peaches. How many peaches does Marin have?

....

3) Seven pears were in the basket. More pears were added to the basket. Now there are 40 pears. How many pears were added to the basket?

....

4) 16 oranges are in the basket. 27 more oranges are put in the basket. How many oranges are in the basket now?

....

5) Adam has 25 plums and Billy has 44 plums. How many plums do Adam and Billy have together?

....

6) 23 red apples and three green apples are in the basket. How many apples are in the basket?

....

7) Some bananas were in the basket. 39 more bananas were added to the basket. Now there are 65 bananas. How many bananas were in the basket before more bananas were added?

....

8) Jake has 17 mangoes and Paul has 21 mangoes. How many mangoes do Jake and Paul have together?

....

SUBTRACTION EXERCISES

1) 39 − 20 = ...
2) 35 − 31 = ...
3) 12 − 11 = ...
4) 43 − 29 = ...
5) 10 − 10 = ...
6) 39 − 13 = ...
7) 39 − 35 = ...
8) 22 − 18 = ...
9) 31 − 29 = ...
10) 29 − 13 = ...
11) 39 − 15 = ...
12) 25 − 12 = ...
13) 28 − 25 = ...
14) 38 − 26 = ...
15) 44 − 32 = ...
16) 26 − 18 = ...
17) 17 − 15 = ...
18) 36 − 12 = ...
19) 42 − 36 = ...
20) 36 − 18 = ...
21) 41 − 24 = ...
22) 21 − 10 = ...
23) 34 − 24 = ...
24) 35 − 14 = ...
25) 42 − 37 = ...
26) 15 − 15 = ...
27) 37 − 14 = ...
28) 20 − 16 = ...
29) 48 − 30 = ...
30) 25 − 13 = ...
31) 11 − 10 = ...
32) 40 − 21 = ...
33) 39 − 17 = ...
34) 47 − 47 = ...
35) 19 − 12 = ...

1) 26 − 21	2) 29 − 24	3) 48 − 35	4) 33 − 10	5) 30 − 10
6) 30 − 20	7) 41 − 36	8) 23 − 16	9) 46 − 28	10) 16 − 10
11) 34 − 17	12) 45 − 42	13) 46 − 31	14) 49 − 13	15) 16 − 11
16) 34 − 24	17) 17 − 10	18) 15 − 13	19) 23 − 14	20) 31 − 19
21) 32 − 16	22) 47 − 25	23) 40 − 40	24) 47 − 13	25) 48 − 45
26) 39 − 13	27) 43 − 27	28) 24 − 12	29) 43 − 29	30) 43 − 23
31) 38 − 24	32) 44 − 19	33) 27 − 26	34) 23 − 20	35) 39 − 28

1) 70 − 42 = ...	2) 39 − 14 = ...	3) 26 − 25 = ...	4) 73 − 41 = ...	5) 28 − 14 = ...
6) 87 − 50 = ...	7) 45 − 20 = ...	8) 10 − 10 = ...	9) 73 − 16 = ...	10) 60 − 19 = ...
11) 95 − 50 = ...	12) 34 − 23 = ...	13) 99 − 84 = ...	14) 58 − 52 = ...	15) 24 − 22 = ...
16) 51 − 37 = ...	17) 34 − 27 = ...	18) 81 − 56 = ...	19) 67 − 58 = ...	20) 19 − 19 = ...
21) 34 − 10 = ...	22) 80 − 15 = ...	23) 98 − 18 = ...	24) 33 − 31 = ...	25) 35 − 27 = ...
26) 76 − 32 = ...	27) 60 − 20 = ...	28) 50 − 18 = ...	29) 33 − 28 = ...	30) 23 − 15 = ...
31) 48 − 42 = ...	32) 40 − 19 = ...	33) 76 − 24 = ...	34) 32 − 25 = ...	35) 53 − 25 = ...

1)	2)	3)	4)	5)
87	94	56	15	31
− 37	− 34	− 33	− 11	− 27
....	....	....	....	....

6)	7)	8)	9)	10)
82	86	64	26	89
− 28	− 22	− 16	− 17	− 85
....	....	....	....	....

11)	12)	13)	14)	15)
78	31	53	29	81
− 22	− 12	− 21	− 26	− 15
....	....	....	....	....

16)	17)	18)	19)	20)
10	68	35	23	76
− 10	− 36	− 30	− 20	− 16
....	....	....	....	....

21)	22)	23)	24)	25)
41	51	73	83	75
− 31	− 49	− 21	− 33	− 53
....	....	....	....	....

26)	27)	28)	29)	30)
68	59	35	20	82
− 17	− 58	− 27	− 18	− 29
....	....	....	....	....

31)	32)	33)	34)	35)
36	45	80	63	89
− 15	− 16	− 54	− 32	− 75
....	....	....	....	....

1)	2)	3)	4)	5)
33 − 20 = ...	78 − 35 = ...	27 − 27 = ...	84 − 44 = ...	55 − 47 = ...

6)	7)	8)	9)	10)
44 − 12 = ...	33 − 27 = ...	91 − 11 = ...	92 − 30 = ...	37 − 14 = ...

11)	12)	13)	14)	15)
69 − 35 = ...	31 − 31 = ...	51 − 42 = ...	46 − 14 = ...	37 − 29 = ...

16)	17)	18)	19)	20)
12 − 11 = ...	31 − 20 = ...	25 − 17 = ...	87 − 34 = ...	96 − 61 = ...

21)	22)	23)	24)	25)
85 − 74 = ...	51 − 36 = ...	47 − 14 = ...	13 − 12 = ...	34 − 12 = ...

26)	27)	28)	29)	30)
69 − 19 = ...	20 − 18 = ...	12 − 12 = ...	26 − 22 = ...	91 − 43 = ...

31)	32)	33)	34)	35)
62 − 48 = ...	23 − 20 = ...	82 − 75 = ...	83 − 68 = ...	61 − 33 = ...

1) 198 − 129 = ...

2) 111 − 12 = ...

3) 175 − 147 = ...

4) 178 − 150 = ...

5) 171 − 122 = ...

6) 177 − 39 = ...

7) 133 − 34 = ...

8) 140 − 76 = ...

9) 119 − 67 = ...

10) 166 − 69 = ...

11) 166 − 66 = ...

12) 190 − 68 = ...

13) 198 − 61 = ...

14) 126 − 77 = ...

15) 101 − 76 = ...

16) 161 − 93 = ...

17) 155 − 61 = ...

18) 180 − 68 = ...

19) 184 − 76 = ...

20) 189 − 154 = ...

21) 164 − 95 = ...

22) 185 − 172 = ...

23) 124 − 88 = ...

24) 140 − 35 = ...

25) 187 − 11 = ...

26) 123 − 67 = ...

27) 123 − 109 = ...

28) 184 − 117 = ...

29) 173 − 137 = ...

30) 103 − 55 = ...

31) 157 − 39 = ...

32) 183 − 180 = ...

33) 109 − 32 = ...

34) 163 − 160 = ...

35) 139 − 17 = ...

1) 142 − 125 = ...

2) 137 − 27 = ...

3) 173 − 27 = ...

4) 175 − 125 = ...

5) 173 − 110 = ...

6) 150 − 78 = ...

7) 183 − 93 = ...

8) 199 − 181 = ...

9) 195 − 96 = ...

10) 120 − 59 = ...

11) 162 − 80 = ...

12) 123 − 80 = ...

13) 130 − 56 = ...

14) 164 − 33 = ...

15) 107 − 33 = ...

16) 145 − 46 = ...

17) 190 − 130 = ...

18) 108 − 32 = ...

19) 195 − 110 = ...

20) 179 − 24 = ...

21) 186 − 72 = ...

22) 161 − 130 = ...

23) 173 − 165 = ...

24) 189 − 79 = ...

25) 184 − 29 = ...

26) 111 − 106 = ...

27) 160 − 66 = ...

28) 172 − 165 = ...

29) 165 − 127 = ...

30) 183 − 69 = ...

31) 146 − 92 = ...

32) 163 − 86 = ...

33) 116 − 35 = ...

34) 159 − 35 = ...

35) 140 − 110 = ...

1) 169 − 53 = ...	2) 148 − 119 = ...	3) 146 − 24 = ...	4) 122 − 107 = ...	5) 147 − 126 = ...
6) 121 − 43 = ...	7) 168 − 68 = ...	8) 166 − 98 = ...	9) 107 − 31 = ...	10) 127 − 32 = ...
11) 176 − 126 = ...	12) 124 − 67 = ...	13) 132 − 76 = ...	14) 173 − 139 = ...	15) 139 − 139 = ...
16) 178 − 54 = ...	17) 123 − 101 = ...	18) 144 − 40 = ...	19) 119 − 90 = ...	20) 149 − 13 = ...
21) 136 − 51 = ...	22) 124 − 79 = ...	23) 141 − 136 = ...	24) 104 − 81 = ...	25) 197 − 184 = ...
26) 129 − 69 = ...	27) 172 − 154 = ...	28) 139 − 138 = ...	29) 104 − 102 = ...	30) 147 − 135 = ...
31) 137 − 64 = ...	32) 148 − 76 = ...	33) 107 − 102 = ...	34) 107 − 94 = ...	35) 193 − 90 = ...

1) 243 − 151 = ...	2) 253 − 170 = ...	3) 214 − 198 = ...	4) 225 − 115 = ...	5) 215 − 124 = ...
6) 130 − 124 = ...	7) 248 − 67 = ...	8) 199 − 88 = ...	9) 166 − 109 = ...	10) 128 − 55 = ...
11) 129 − 47 = ...	12) 274 − 107 = ...	13) 293 − 161 = ...	14) 170 − 158 = ...	15) 160 − 96 = ...
16) 113 − 82 = ...	17) 277 − 162 = ...	18) 221 − 172 = ...	19) 111 − 25 = ...	20) 284 − 177 = ...
21) 154 − 101 = ...	22) 211 − 107 = ...	23) 283 − 113 = ...	24) 159 − 47 = ...	25) 213 − 127 = ...
26) 234 − 23 = ...	27) 270 − 26 = ...	28) 219 − 86 = ...	29) 274 − 58 = ...	30) 275 − 157 = ...
31) 228 − 142 = ...	32) 150 − 11 = ...	33) 207 − 96 = ...	34) 113 − 107 = ...	35) 239 − 132 = ...

1) 198 − 76 = ...

2) 217 − 194 = ...

3) 116 − 78 = ...

4) 202 − 134 = ...

5) 190 − 59 = ...

6) 207 − 74 = ...

7) 214 − 30 = ...

8) 208 − 124 = ...

9) 153 − 115 = ...

10) 198 − 134 = ...

11) 210 − 182 = ...

12) 214 − 15 = ...

13) 209 − 35 = ...

14) 142 − 61 = ...

15) 127 − 121 = ...

16) 130 − 77 = ...

17) 187 − 84 = ...

18) 182 − 121 = ...

19) 153 − 129 = ...

20) 125 − 75 = ...

21) 194 − 38 = ...

22) 265 − 175 = ...

23) 241 − 55 = ...

24) 262 − 95 = ...

25) 188 − 163 = ...

26) 142 − 92 = ...

27) 112 − 37 = ...

28) 242 − 122 = ...

29) 256 − 117 = ...

30) 185 − 176 = ...

31) 187 − 32 = ...

32) 203 − 105 = ...

33) 125 − 123 = ...

34) 280 − 38 = ...

35) 239 − 180 = ...

1) 44 apples are in the basket. 43 apples are taken out of the basket. How many apples are in the basket now?

....

2) 69 peaches were in the basket. Some of the peaches were removed from the basket. Now there are 19 peaches. How many peaches were removed from the basket?

....

3) 72 plums are in the basket. 45 are red and the rest are green. How many plums are green?

....

4) Paul has 13 bananas. Brian has 19 bananas. How many more bananas does Brian have than Paul?

....

5) Michele has 50 fewer oranges than Marin. Marin has 66 oranges. How many oranges does Michele have?

....

6) Some pears were in the basket. 27 pears were taken from the basket. Now there are zero pears. How many pears were in the basket before some of the pears were taken?

....

7) Paul has 41 mangoes. Brian has 43 mangoes. How many more mangoes does Brian have than Paul?

....

8) 13 plums were in the basket. Some of the plums were removed from the basket. Now there is one plum. How many plums were removed from the basket?

....

1) Some apples were in the basket. 26 apples were taken from the basket. Now there are six apples. How many apples were in the basket before some of the apples were taken?

....

2) 11 peaches are in the basket. 11 are red and the rest are green. How many peaches are green?

....

3) Allan has 29 mangoes. Steven has 34 mangoes. How many more mangoes does Steven have than Allan?

....

4) 94 plums were in the basket. Some of the plums were removed from the basket. Now there are 52 plums. How many plums were removed from the basket?

....

5) Amy has 65 fewer oranges than Janet. Janet has 92 oranges. How many oranges does Amy have?

....

6) 20 pears are in the basket. 10 pears are taken out of the basket. How many pears are in the basket now?

....

7) Marcie has 22 fewer bananas than Marin. Marin has 62 bananas. How many bananas does Marcie have?

....

8) Brian has 38 plums. Billy has 54 plums. How many more plums does Billy have than Brian?

....

1) Jackie has 39 fewer plums than Marin. Marin has 55 plums. How many plums does Jackie have?

2) Steven has 32 bananas. Jake has 57 bananas. How many more bananas does Jake have than Steven?

3) 51 peaches were in the basket. Some of the peaches were removed from the basket. Now there are 25 peaches. How many peaches were removed from the basket?

4) Some apples were in the basket. 35 apples were taken from the basket. Now there are six apples. How many apples were in the basket before some of the apples were taken?

5) 32 pears are in the basket. 31 are red and the rest are green. How many pears are green?

6) 26 mangoes are in the basket. 25 mangoes are taken out of the basket. How many mangoes are in the basket now?

7) Allan has 43 oranges. Paul has 62 oranges. How many more oranges does Paul have than Allan?

8) Some pears were in the basket. 36 pears were taken from the basket. Now there are six pears. How many pears were in the basket before some of the pears were taken?

1) 61 apples were in the basket. Some of the apples were removed from the basket. Now there are 50 apples. How many apples were removed from the basket?

2) 86 oranges are in the basket. 22 are red and the rest are green. How many oranges are green?

3) Amy has 50 fewer pears than Jackie. Jackie has 88 pears. How many pears does Amy have?

4) Some peaches were in the basket. 11 peaches were taken from the basket. Now there are three peaches. How many peaches were in the basket before some of the peaches were taken?

5) 58 mangoes are in the basket. 24 mangoes are taken out of the basket. How many mangoes are in the basket now?

6) Donald has 24 bananas. Jake has 29 bananas. How many more bananas does Jake have than Donald?

7) Marin has seven fewer plums than Marcie. Marcie has 46 plums. How many plums does Marin have?

8) Some bananas were in the basket. 25 bananas were taken from the basket. Now there are 41 bananas. How many bananas were in the basket before some of the bananas were taken?

MULTIPLICATION EXERCISES

1) 4×9	2) 5×7	3) 7×3	4) 7×6	5) 8×4 ....
6) 2×3	7) 7×8	8) 3×9	9) 4×6	10) 5×3 ....
11) 5×6	12) 5×8	13) 6×5	14) 3×4	15) 8×10 ....
16) 8×6	17) 8×7	18) 10×4	19) 3×8	20) 5×10 ....
21) 8×3	22) 3×3	23) 4×2	24) 8×9	25) 9×7 ....
26) 10×5	27) 8×8	28) 7×5	29) 7×4	30) 6×3 ....
31) 10×7	32) 6×7	33) 9×3	34) 7×10	35) 3×6 ....

1) $6 \times 7 =$ ....
2) $7 \times 3 =$ ....
3) $5 \times 2 =$ ....
4) $5 \times 4 =$ ....
5) $3 \times 7 =$ ....
6) $7 \times 4 =$ ....
7) $10 \times 6 =$ ....
8) $6 \times 4 =$ ....
9) $4 \times 6 =$ ....
10) $9 \times 9 =$ ....
11) $8 \times 7 =$ ....
12) $4 \times 10 =$ ....
13) $7 \times 10 =$ ....
14) $9 \times 8 =$ ....
15) $6 \times 9 =$ ....
16) $8 \times 6 =$ ....
17) $3 \times 10 =$ ....
18) $5 \times 8 =$ ....
19) $2 \times 9 =$ ....
20) $6 \times 6 =$ ....
21) $2 \times 6 =$ ....
22) $6 \times 8 =$ ....
23) $4 \times 5 =$ ....
24) $9 \times 6 =$ ....
25) $2 \times 5 =$ ....
26) $8 \times 9 =$ ....
27) $10 \times 8 =$ ....
28) $8 \times 8 =$ ....
29) $8 \times 4 =$ ....
30) $7 \times 6 =$ ....
31) $9 \times 2 =$ ....
32) $10 \times 3 =$ ....
33) $7 \times 7 =$ ....
34) $7 \times 9 =$ ....
35) $9 \times 10 =$ ....

1) 9×7 = ...

2) 47×5 = ...

3) 20×4 = ...

4) 45×6 = ...

5) 34×7 = ...

6) 20×9 = ...

7) 47×4 = ...

8) 9×5 = ...

9) 13×3 = ...

10) 19×9 = ...

11) 33×6 = ...

12) 19×5 = ...

13) 27×4 = ...

14) 39×9 = ...

15) 28×3 = ...

16) 21×7 = ...

17) 34×3 = ...

18) 20×3 = ...

19) 5×6 = ...

20) 14×6 = ...

21) 4×8 = ...

22) 45×10 = ...

23) 32×10 = ...

24) 34×5 = ...

25) 48×4 = ...

26) 15×8 = ...

27) 23×8 = ...

28) 17×6 = ...

29) 43×3 = ...

30) 23×3 = ...

31) 46×7 = ...

32) 12×8 = ...

33) 10×9 = ...

34) 44×3 = ...

35) 16×7 = ...

1) $38 \times 3 =$ ....

2) $13 \times 4 =$ ....

3) $38 \times 10 =$ ....

4) $16 \times 6 =$ ....

5) $33 \times 9 =$ ....

6) $40 \times 6 =$ ....

7) $50 \times 5 =$ ....

8) $16 \times 4 =$ ....

9) $12 \times 4 =$ ....

10) $33 \times 3 =$ ....

11) $22 \times 7 =$ ....

12) $48 \times 10 =$ ....

13) $18 \times 8 =$ ....

14) $44 \times 7 =$ ....

15) $29 \times 2 =$ ....

16) $42 \times 5 =$ ....

17) $33 \times 7 =$ ....

18) $21 \times 5 =$ ....

19) $2 \times 7 =$ ....

20) $21 \times 3 =$ ....

21) $38 \times 4 =$ ....

22) $37 \times 2 =$ ....

23) $31 \times 9 =$ ....

24) $26 \times 6 =$ ....

25) $46 \times 4 =$ ....

26) $20 \times 4 =$ ....

27) $13 \times 7 =$ ....

28) $38 \times 7 =$ ....

29) $17 \times 8 =$ ....

30) $33 \times 6 =$ ....

31) $31 \times 10 =$ ....

32) $19 \times 8 =$ ....

33) $32 \times 7 =$ ....

34) $35 \times 7 =$ ....

35) $43 \times 8 =$ ....

1) 49×6 = ...	2) 32×3 = ...	3) 75×7 = ...	4) 93×9 = ...	5) 34×10 = ...
6) 52×6 = ...	7) 77×3 = ...	8) 20×3 = ...	9) 79×4 = ...	10) 71×6 = ...
11) 91×3 = ...	12) 90×7 = ...	13) 78×8 = ...	14) 83×10 = ...	15) 71×8 = ...
16) 87×6 = ...	17) 27×9 = ...	18) 13×3 = ...	19) 88×5 = ...	20) 33×8 = ...
21) 28×4 = ...	22) 49×8 = ...	23) 63×3 = ...	24) 24×8 = ...	25) 73×7 = ...
26) 79×2 = ...	27) 21×4 = ...	28) 77×4 = ...	29) 65×7 = ...	30) 46×2 = ...
31) 79×10 = ...	32) 25×6 = ...	33) 16×10 = ...	34) 36×9 = ...	35) 22×2 = ...

1) $40 \times 2 =$ ....

2) $29 \times 2 =$ ....

3) $15 \times 6 =$ ....

4) $55 \times 7 =$ ....

5) $76 \times 6 =$ ....

6) $70 \times 8 =$ ....

7) $16 \times 7 =$ ....

8) $32 \times 6 =$ ....

9) $44 \times 8 =$ ....

10) $43 \times 6 =$ ....

11) $97 \times 6 =$ ....

12) $79 \times 9 =$ ....

13) $14 \times 7 =$ ....

14) $84 \times 5 =$ ....

15) $41 \times 9 =$ ....

16) $58 \times 8 =$ ....

17) $65 \times 8 =$ ....

18) $27 \times 9 =$ ....

19) $18 \times 4 =$ ....

20) $97 \times 7 =$ ....

21) $27 \times 8 =$ ....

22) $39 \times 6 =$ ....

23) $98 \times 4 =$ ....

24) $52 \times 3 =$ ....

25) $49 \times 7 =$ ....

26) $28 \times 3 =$ ....

27) $47 \times 9 =$ ....

28) $32 \times 4 =$ ....

29) $96 \times 5 =$ ....

30) $73 \times 3 =$ ....

31) $37 \times 9 =$ ....

32) $29 \times 5 =$ ....

33) $96 \times 6 =$ ....

34) $41 \times 6 =$ ....

35) $64 \times 3 =$ ....

1) $18 \times 5 =$ ....

2) $24 \times 3 =$ ....

3) $182 \times 10 =$ ....

4) $58 \times 6 =$ ....

5) $51 \times 3 =$ ....

6) $63 \times 10 =$ ....

7) $99 \times 6 =$ ....

8) $188 \times 6 =$ ....

9) $150 \times 3 =$ ....

10) $82 \times 3 =$ ....

11) $61 \times 5 =$ ....

12) $117 \times 9 =$ ....

13) $199 \times 6 =$ ....

14) $70 \times 6 =$ ....

15) $74 \times 4 =$ ....

16) $98 \times 7 =$ ....

17) $153 \times 5 =$ ....

18) $28 \times 8 =$ ....

19) $95 \times 8 =$ ....

20) $200 \times 5 =$ ....

21) $116 \times 9 =$ ....

22) $138 \times 4 =$ ....

23) $181 \times 5 =$ ....

24) $193 \times 3 =$ ....

25) $170 \times 7 =$ ....

26) $17 \times 5 =$ ....

27) $90 \times 4 =$ ....

28) $50 \times 7 =$ ....

29) $157 \times 4 =$ ....

30) $34 \times 5 =$ ....

31) $136 \times 8 =$ ....

32) $46 \times 9 =$ ....

33) $145 \times 7 =$ ....

34) $55 \times 3 =$ ....

35) $80 \times 3 =$ ....

1) 12×4 = ...

2) 97×7 = ...

3) 160×10 = ...

4) 109×4 = ...

5) 98×8 = ...

6) 102×4 = ...

7) 94×7 = ...

8) 118×5 = ...

9) 138×2 = ...

10) 187×4 = ...

11) 162×6 = ...

12) 96×4 = ...

13) 156×8 = ...

14) 191×7 = ...

15) 136×7 = ...

16) 159×5 = ...

17) 191×3 = ...

18) 58×9 = ...

19) 157×9 = ...

20) 70×6 = ...

21) 97×9 = ...

22) 34×9 = ...

23) 62×7 = ...

24) 51×3 = ...

25) 16×9 = ...

26) 90×8 = ...

27) 169×6 = ...

28) 158×7 = ...

29) 154×4 = ...

30) 180×2 = ...

31) 199×9 = ...

32) 132×9 = ...

33) 193×3 = ...

34) 56×4 = ...

35) 86×7 = ...

1) $168 \times 5 =$ ....

2) $127 \times 5 =$ ....

3) $197 \times 9 =$ ....

4) $173 \times 10 =$ ....

5) $155 \times 10 =$ ....

6) $156 \times 6 =$ ....

7) $132 \times 9 =$ ....

8) $171 \times 9 =$ ....

9) $165 \times 7 =$ ....

10) $157 \times 7 =$ ....

11) $112 \times 2 =$ ....

12) $114 \times 4 =$ ....

13) $195 \times 8 =$ ....

14) $127 \times 6 =$ ....

15) $116 \times 5 =$ ....

16) $159 \times 7 =$ ....

17) $164 \times 9 =$ ....

18) $148 \times 9 =$ ....

19) $111 \times 3 =$ ....

20) $140 \times 9 =$ ....

21) $154 \times 4 =$ ....

22) $169 \times 5 =$ ....

23) $109 \times 10 =$ ....

24) $159 \times 6 =$ ....

25) $164 \times 3 =$ ....

26) $156 \times 5 =$ ....

27) $184 \times 10 =$ ....

28) $153 \times 7 =$ ....

29) $130 \times 2 =$ ....

30) $148 \times 4 =$ ....

31) $159 \times 4 =$ ....

32) $182 \times 3 =$ ....

33) $141 \times 10 =$ ....

34) $189 \times 5 =$ ....

35) $138 \times 9 =$ ....

1) $109 \times 4 =$ ....

2) $115 \times 9 =$ ....

3) $170 \times 4 =$ ....

4) $152 \times 9 =$ ....

5) $150 \times 3 =$ ....

6) $179 \times 6 =$ ....

7) $182 \times 4 =$ ....

8) $124 \times 4 =$ ....

9) $124 \times 2 =$ ....

10) $152 \times 4 =$ ....

11) $175 \times 8 =$ ....

12) $106 \times 5 =$ ....

13) $186 \times 5 =$ ....

14) $182 \times 9 =$ ....

15) $199 \times 9 =$ ....

16) $196 \times 8 =$ ....

17) $102 \times 7 =$ ....

18) $177 \times 10 =$ ....

19) $161 \times 8 =$ ....

20) $168 \times 7 =$ ....

21) $104 \times 3 =$ ....

22) $121 \times 2 =$ ....

23) $150 \times 9 =$ ....

24) $164 \times 4 =$ ....

25) $171 \times 10 =$ ....

26) $185 \times 3 =$ ....

27) $158 \times 6 =$ ....

28) $160 \times 5 =$ ....

29) $114 \times 8 =$ ....

30) $100 \times 5 =$ ....

31) $144 \times 8 =$ ....

32) $192 \times 7 =$ ....

33) $115 \times 3 =$ ....

34) $143 \times 3 =$ ....

35) $153 \times 4 =$ ....

1) If there are 5 bananas in each box and there are 5 boxes, how many bananas are there in total?

....

2) Steven can cycle 7 miles per hour. How far can Steven cycle in 9 hours?

....

3) Paul has 19 times more oranges than Janet. Janet has 6 oranges. How many oranges does Paul have?

....

4) David swims 18 laps every day. How many laps will David swim in 3 days?

....

5) Amy's garden has 13 rows of pumpkins. Each row has 5 pumpkins. How many pumpkins does Amy have in all?

....

6) If there are 20 peaches in each box and there are 3 boxes, how many peaches are there in total?

....

7) Audrey has 11 times more mangoes than Sharon. Sharon has 7 mangoes. How many mangoes does Audrey have?

....

8) Michele's garden has 19 rows of pumpkins. Each row has 2 pumpkins. How many pumpkins does Michele have in all?

....

1) Janet has 20 times more peaches than David. David has 7 peaches. How many peaches does Janet have?

2) If there are 35 oranges in each box and there are 2 boxes, how many oranges are there in total?

3) Brian can cycle 3 miles per hour. How far can Brian cycle in 2 hours?

4) Amy's garden has 43 rows of pumpkins. Each row has 3 pumpkins. How many pumpkins does Amy have in all?

5) Jackie swims 10 laps every day. How many laps will Jackie swim in 5 days?

6) If there are 33 bananas in each box and there are 3 boxes, how many bananas are there in total?

7) Adam swims 40 laps every day. How many laps will Adam swim in 3 days?

8) Sharon's garden has 36 rows of pumpkins. Each row has 9 pumpkins. How many pumpkins does Sharon have in all?

1) Adam has 43 times more peaches than Brian. Brian has 4 peaches. How many peaches does Adam have?

....

2) Billy can cycle 6 miles per hour. How far can Billy cycle in 3 hours?

....

3) If there are 28 oranges in each box and there are 7 boxes, how many oranges are there in total?

....

4) Steven swims 12 laps every day. How many laps will Steven swim in 6 days?

....

5) Jackie's garden has 39 rows of pumpkins. Each row has 3 pumpkins. How many pumpkins does Jackie have in all?

....

6) Jake swims 13 laps every day. How many laps will Jake swim in 6 days?

....

7) Jake can cycle 17 miles per hour. How far can Jake cycle in 9 hours?

....

8) Amy's garden has 11 rows of pumpkins. Each row has 10 pumpkins. How many pumpkins does Amy have in all?

....

1) Ellen's garden has 18 rows of pumpkins. Each row has 2 pumpkins. How many pumpkins does Ellen have in all?

....

2) Jackie has 41 times more apples than Janet. Janet has 9 apples. How many apples does Jackie have?

....

3) If there are 47 plums in each box and there are 9 boxes, how many plums are there in total?

....

4) Adam swims 5 laps every day. How many laps will Adam swim in 5 days?

....

5) Steven can cycle 41 miles per hour. How far can Steven cycle in 10 hours?

....

6) If there are 3 peaches in each box and there are 9 boxes, how many peaches are there in total?

....

7) Steven has 3 times more bananas than Ellen. Ellen has 6 bananas. How many bananas does Steven have?

....

8) Allan can cycle 16 miles per hour. How far can Allan cycle in 5 hours?

....

DIVISION EXERCISES

1) $25 \div 5 =$ ....
2) $4 \div 2 =$ ....
3) $56 \div 8 =$ ....
4) $6 \div 3 =$ ....
5) $32 \div 4 =$ ....
6) $27 \div 9 =$ ....
7) $48 \div 8 =$ ....
8) $48 \div 6 =$ ....
9) $54 \div 9 =$ ....
10) $40 \div 10 =$ ....
11) $24 \div 3 =$ ....
12) $3 \div 3 =$ ....
13) $27 \div 3 =$ ....
14) $42 \div 6 =$ ....
15) $49 \div 7 =$ ....
16) $40 \div 5 =$ ....
17) $7 \div 7 =$ ....
18) $35 \div 7 =$ ....
19) $18 \div 3 =$ ....
20) $50 \div 10 =$ ....
21) $36 \div 9 =$ ....
22) $72 \div 8 =$ ....
23) $6 \div 6 =$ ....
24) $30 \div 10 =$ ....
25) $20 \div 4 =$ ....
26) $5 \div 5 =$ ....
27) $70 \div 10 =$ ....
28) $16 \div 4 =$ ....
29) $8 \div 4 =$ ....
30) $72 \div 9 =$ ....
31) $10 \div 2 =$ ....
32) $24 \div 6 =$ ....
33) $12 \div 4 =$ ....
34) $21 \div 3 =$ ....
35) $16 \div 8 =$ ....
36) $12 \div 6 =$ ....

1) $18 \div 9 =$ ....
2) $15 \div 3 =$ ....
3) $12 \div 4 =$ ....
4) $20 \div 5 =$ ....
5) $21 \div 7 =$ ....
6) $4 \div 4 =$ ....
7) $10 \div 2 =$ ....
8) $54 \div 6 =$ ....
9) $45 \div 9 =$ ....
10) $14 \div 7 =$ ....
11) $4 \div 2 =$ ....
12) $32 \div 4 =$ ....
13) $9 \div 9 =$ ....
14) $36 \div 6 =$ ....
15) $36 \div 9 =$ ....
16) $30 \div 5 =$ ....
17) $18 \div 2 =$ ....
18) $35 \div 7 =$ ....
19) $16 \div 4 =$ ....
20) $64 \div 8 =$ ....
21) $48 \div 6 =$ ....
22) $18 \div 3 =$ ....
23) $12 \div 3 =$ ....
24) $12 \div 2 =$ ....
25) $27 \div 3 =$ ....
26) $24 \div 3 =$ ....
27) $28 \div 7 =$ ....
28) $28 \div 4 =$ ....
29) $25 \div 5 =$ ....
30) $81 \div 9 =$ ....
31) $24 \div 6 =$ ....
32) $72 \div 9 =$ ....
33) $56 \div 8 =$ ....
34) $3 \div 3 =$ ....
35) $42 \div 6 =$ ....
36) $60 \div 10 =$ ....

1) 10 ÷ 2 = ...

2) 28 ÷ 4 = ...

3) 9 ÷ 3 = ...

4) 20 ÷ 4 = ...

5) 12 ÷ 2 = ...

6) 42 ÷ 6 = ...

7) 27 ÷ 9 = ...

8) 28 ÷ 7 = ...

9) 25 ÷ 5 = ...

10) 6 ÷ 6 = ...

11) 18 ÷ 3 = ...

12) 32 ÷ 4 = ...

13) 40 ÷ 8 = ...

14) 35 ÷ 5 = ...

15) 30 ÷ 5 = ...

16) 20 ÷ 10 = ...

17) 64 ÷ 8 = ...

18) 45 ÷ 9 = ...

19) 12 ÷ 6 = ...

20) 63 ÷ 9 = ...

21) 12 ÷ 3 = ...

22) 27 ÷ 3 = ...

23) 24 ÷ 6 = ...

24) 16 ÷ 4 = ...

25) 21 ÷ 3 = ...

26) 54 ÷ 9 = ...

27) 63 ÷ 7 = ...

28) 7 ÷ 7 = ...

29) 72 ÷ 9 = ...

30) 56 ÷ 7 = ...

31) 35 ÷ 7 = ...

32) 21 ÷ 7 = ...

33) 70 ÷ 10 = ...

34) 50 ÷ 10 = ...

35) 4 ÷ 2 = ...

36) 15 ÷ 5 = ...

1) $16 \div 8 =$ ....
2) $7 \div 7 =$ ....
3) $27 \div 9 =$ ....
4) $14 \div 7 =$ ....
5) $36 \div 6 =$ ....
6) $8 \div 8 =$ ....
7) $72 \div 8 =$ ....
8) $6 \div 2 =$ ....
9) $12 \div 3 =$ ....
10) $35 \div 7 =$ ....
11) $42 \div 7 =$ ....
12) $12 \div 6 =$ ....
13) $45 \div 9 =$ ....
14) $30 \div 5 =$ ....
15) $21 \div 3 =$ ....
16) $10 \div 5 =$ ....
17) $18 \div 6 =$ ....
18) $9 \div 3 =$ ....
19) $32 \div 4 =$ ....
20) $24 \div 6 =$ ....
21) $8 \div 4 =$ ....
22) $2 \div 2 =$ ....
23) $15 \div 3 =$ ....
24) $63 \div 7 =$ ....
25) $12 \div 4 =$ ....
26) $6 \div 6 =$ ....
27) $42 \div 6 =$ ....
28) $56 \div 8 =$ ....
29) $48 \div 6 =$ ....
30) $48 \div 8 =$ ....
31) $18 \div 3 =$ ....
32) $20 \div 10 =$ ....
33) $25 \div 5 =$ ....
34) $40 \div 5 =$ ....
35) $4 \div 2 =$ ....
36) $28 \div 7 =$ ....

1) $18 \div 3 =$ ....
2) $2 \div 2 =$ ....
3) $6 \div 6 =$ ....

4) $45 \div 9 =$ ....
5) $35 \div 5 =$ ....
6) $42 \div 7 =$ ....

7) $20 \div 5 =$ ....
8) $49 \div 7 =$ ....
9) $21 \div 7 =$ ....

10) $56 \div 8 =$ ....
11) $63 \div 7 =$ ....
12) $16 \div 4 =$ ....

13) $8 \div 2 =$ ....
14) $24 \div 4 =$ ....
15) $36 \div 4 =$ ....

16) $27 \div 3 =$ ....
17) $21 \div 3 =$ ....
18) $42 \div 6 =$ ....

19) $28 \div 4 =$ ....
20) $64 \div 8 =$ ....
21) $36 \div 9 =$ ....

22) $12 \div 3 =$ ....
23) $70 \div 10 =$ ....
24) $35 \div 7 =$ ....

25) $4 \div 2 =$ ....
26) $16 \div 8 =$ ....
27) $14 \div 2 =$ ....

28) $15 \div 5 =$ ....
29) $32 \div 4 =$ ....
30) $12 \div 2 =$ ....

31) $60 \div 10 =$ ....
32) $40 \div 8 =$ ....
33) $12 \div 6 =$ ....

34) $30 \div 10 =$ ....
35) $25 \div 5 =$ ....
36) $48 \div 8 =$ ....

1) $30 \div 6 =$ ....
2) $8 \div 2 =$ ....
3) $30 \div 15 =$ ....
4) $68 \div 17 =$ ....
5) $60 \div 12 =$ ....
6) $28 \div 4 =$ ....
7) $49 \div 7 =$ ....
8) $34 \div 17 =$ ....
9) $10 \div 5 =$ ....
10) $112 \div 14 =$ ....
11) $30 \div 10 =$ ....
12) $128 \div 16 =$ ....
13) $32 \div 4 =$ ....
14) $4 \div 4 =$ ....
15) $38 \div 19 =$ ....
16) $95 \div 19 =$ ....
17) $48 \div 16 =$ ....
18) $84 \div 14 =$ ....
19) $7 \div 7 =$ ....
20) $8 \div 4 =$ ....
21) $54 \div 9 =$ ....
22) $28 \div 14 =$ ....
23) $21 \div 7 =$ ....
24) $85 \div 17 =$ ....
25) $80 \div 20 =$ ....
26) $20 \div 4 =$ ....
27) $9 \div 3 =$ ....
28) $45 \div 5 =$ ....
29) $45 \div 15 =$ ....
30) $56 \div 7 =$ ....
31) $133 \div 19 =$ ....
32) $40 \div 10 =$ ....
33) $66 \div 11 =$ ....
34) $2 \div 2 =$ ....
35) $6 \div 3 =$ ....
36) $36 \div 18 =$ ....

1) $60 \div 15 =$	2) $18 \div 9 =$	3) $56 \div 14 =$ ....
4) $72 \div 18 =$	5) $120 \div 15 =$	6) $81 \div 9 =$ ....
7) $35 \div 7 =$	8) $76 \div 19 =$	9) $42 \div 6 =$ ....
10) $72 \div 9 =$	11) $48 \div 6 =$	12) $152 \div 19 =$ ....
13) $45 \div 15 =$	14) $12 \div 4 =$	15) $8 \div 4 =$ ....
16) $21 \div 3 =$	17) $119 \div 17 =$	18) $56 \div 7 =$ ....
19) $40 \div 10 =$	20) $54 \div 18 =$	21) $48 \div 8 =$ ....
22) $32 \div 16 =$	23) $84 \div 12 =$	24) $72 \div 12 =$ ....
25) $14 \div 14 =$	26) $28 \div 4 =$	27) $84 \div 14 =$ ....
28) $28 \div 7 =$	29) $64 \div 16 =$	30) $95 \div 19 =$ ....
31) $42 \div 14 =$	32) $12 \div 3 =$	33) $16 \div 8 =$ ....
34) $48 \div 12 =$	35) $8 \div 2 =$	36) $36 \div 12 =$ ....

1) $28 \div 7 =$ ....
2) $60 \div 15 =$ ....
3) $85 \div 17 =$ ....
4) $28 \div 4 =$ ....
5) $78 \div 13 =$ ....
6) $81 \div 9 =$ ....
7) $96 \div 16 =$ ....
8) $36 \div 18 =$ ....
9) $77 \div 11 =$ ....
10) $24 \div 3 =$ ....
11) $38 \div 19 =$ ....
12) $36 \div 9 =$ ....
13) $32 \div 4 =$ ....
14) $128 \div 16 =$ ....
15) $22 \div 11 =$ ....
16) $24 \div 8 =$ ....
17) $30 \div 15 =$ ....
18) $42 \div 14 =$ ....
19) $45 \div 15 =$ ....
20) $40 \div 10 =$ ....
21) $91 \div 13 =$ ....
22) $6 \div 3 =$ ....
23) $18 \div 6 =$ ....
24) $45 \div 9 =$ ....
25) $48 \div 12 =$ ....
26) $33 \div 11 =$ ....
27) $70 \div 10 =$ ....
28) $133 \div 19 =$ ....
29) $16 \div 4 =$ ....
30) $65 \div 13 =$ ....
31) $136 \div 17 =$ ....
32) $20 \div 10 =$ ....
33) $14 \div 2 =$ ....
34) $9 \div 3 =$ ....
35) $35 \div 5 =$ ....
36) $44 \div 11 =$ ....

1) 175 ÷ 25 = ...
2) 161 ÷ 23 = ...
3) 144 ÷ 24 = ...
4) 171 ÷ 19 = ...
5) 84 ÷ 21 = ...
6) 133 ÷ 19 = ...
7) 90 ÷ 15 = ...
8) 48 ÷ 16 = ...
9) 76 ÷ 19 = ...
10) 77 ÷ 11 = ...
11) 52 ÷ 13 = ...
12) 30 ÷ 10 = ...
13) 135 ÷ 15 = ...
14) 154 ÷ 22 = ...
15) 72 ÷ 12 = ...
16) 240 ÷ 30 = ...
17) 65 ÷ 13 = ...
18) 38 ÷ 19 = ...
19) 153 ÷ 17 = ...
20) 56 ÷ 28 = ...
21) 48 ÷ 24 = ...
22) 23 ÷ 23 = ...
23) 68 ÷ 17 = ...
24) 84 ÷ 12 = ...
25) 140 ÷ 28 = ...
26) 98 ÷ 14 = ...
27) 36 ÷ 18 = ...
28) 44 ÷ 11 = ...
29) 96 ÷ 24 = ...
30) 147 ÷ 21 = ...
31) 174 ÷ 29 = ...
32) 216 ÷ 27 = ...
33) 102 ÷ 17 = ...
34) 30 ÷ 15 = ...
35) 145 ÷ 29 = ...
36) 152 ÷ 19 = ...

1) 154 ÷ 22 = ...
2) 133 ÷ 19 = ...
3) 54 ÷ 27 = ...
4) 78 ÷ 26 = ...
5) 36 ÷ 18 = ...
6) 60 ÷ 12 = ...
7) 34 ÷ 17 = ...
8) 192 ÷ 24 = ...
9) 39 ÷ 13 = ...
10) 23 ÷ 23 = ...
11) 32 ÷ 16 = ...
12) 55 ÷ 11 = ...
13) 203 ÷ 29 = ...
14) 232 ÷ 29 = ...
15) 40 ÷ 10 = ...
16) 140 ÷ 20 = ...
17) 126 ÷ 21 = ...
18) 17 ÷ 17 = ...
19) 14 ÷ 14 = ...
20) 102 ÷ 17 = ...
21) 69 ÷ 23 = ...
22) 120 ÷ 30 = ...
23) 84 ÷ 12 = ...
24) 21 ÷ 21 = ...
25) 243 ÷ 27 = ...
26) 114 ÷ 19 = ...
27) 54 ÷ 18 = ...
28) 72 ÷ 24 = ...
29) 196 ÷ 28 = ...
30) 24 ÷ 12 = ...
31) 189 ÷ 21 = ...
32) 184 ÷ 23 = ...
33) 90 ÷ 18 = ...
34) 174 ÷ 29 = ...
35) 162 ÷ 27 = ...
36) 161 ÷ 23 = ...

1) Michele made 160 cookies for a bake sale. She put the cookies in bags, with 20 cookies in each bag. How many bags did she have for the bake sale?

....

2) Sandra ordered nine pizzas. The bill for the pizzas came to $126. What was the cost of each pizza?

....

3) A box of bananas weighs 360 pounds. If one banana weighs 20 pounds, how many bananas are there in the box?

....

4) How many 13 cm pieces of rope can you cut from a rope that is 195 cm long?

....

5) You have 70 plums and want to share them equally with 14 people. How many plums would each person get?

....

6) Brian is reading a book with 170 pages. If Brian wants to read the same number of pages every day, how many pages would Brian have to read each day to finish in 10 days?

....

7) Steven is reading a book with 175 pages. If Steven wants to read the same number of pages every day, how many pages would Steven have to read each day to finish in 25 days?

....

8) How many 26 cm pieces of rope can you cut from a rope that is 234 cm long?

....

1) A box of oranges weighs 105 pounds. If one orange weighs seven pounds, how many oranges are there in the box?

....

2) You have 416 pears and want to share them equally with 26 people. How many pears would each person get?

....

3) How many 12 cm pieces of rope can you cut from a rope that is 216 cm long?

....

4) Billy is reading a book with 560 pages. If Billy wants to read the same number of pages every day, how many pages would Billy have to read each day to finish in 28 days?

....

5) Steven ordered eight pizzas. The bill for the pizzas came to $112. What was the cost of each pizza?

....

6) Jackie made 143 cookies for a bake sale. She put the cookies in bags, with 11 cookies in each bag. How many bags did she have for the bake sale?

....

7) Billy is reading a book with 180 pages. If Billy wants to read the same number of pages every day, how many pages would Billy have to read each day to finish in 20 days?

....

8) Amy ordered 15 pizzas. The bill for the pizzas came to $105. What was the cost of each pizza?

....

1) David is reading a book with 459 pages. If David wants to read the same number of pages every day, how many pages would David have to read each day to finish in 27 days?

2) Marin ordered 27 pizzas. The bill for the pizzas came to $540. What was the cost of each pizza?

3) How many 17 cm pieces of rope can you cut from a rope that is 204 cm long?

4) You have 110 mangoes and want to share them equally with 10 people. How many mangoes would each person get?

5) Sharon made 57 cookies for a bake sale. She put the cookies in bags, with 19 cookies in each bag. How many bags did she have for the bake sale?

6) A box of peaches weighs 208 pounds. If one peach weighs 13 pounds, how many peaches are there in the box?

7) Jackie ordered 17 pizzas. The bill for the pizzas came to $102. What was the cost of each pizza?

8) You have 209 plums and want to share them equally with 19 people. How many plums would each person get?

1) A box of pears weighs 39 pounds. If one pear weighs three pounds, how many pears are there in the box?

....

2) How many 20 cm pieces of rope can you cut from a rope that is 140 cm long?

....

3) Marin ordered 19 pizzas. The bill for the pizzas came to $228. What was the cost of each pizza?

....

4) Allan is reading a book with 51 pages. If Allan wants to read the same number of pages every day, how many pages would Allan have to read each day to finish in three days?

....

5) You have 90 apples and want to share them equally with 15 people. How many apples would each person get?

....

6) Marcie made 266 cookies for a bake sale. She put the cookies in bags, with 14 cookies in each bag. How many bags did she have for the bake sale?

....

7) A box of mangoes weighs 256 pounds. If one mango weighs 16 pounds, how many mangoes are there in the box?

....

8) Allan is reading a book with eight pages. If Allan wants to read the same number of pages every day, how many pages would Allan have to read each day to finish in four days?

....

ANSWERS

ADDITION Page 1

1) 26 + 57 = 83	2) 63 + 99 = 162	3) 50 + 45 = 95	4) 65 + 73 = 138	5) 15 + 62 = 77
6) 37 + 41 = 78	7) 49 + 11 = 60	8) 88 + 28 = 116	9) 18 + 47 = 65	10) 21 + 49 = 70
11) 12 + 21 = 33	12) 55 + 79 = 134	13) 79 + 21 = 100	14) 95 + 46 = 141	15) 97 + 68 = 165
16) 32 + 65 = 97	17) 94 + 12 = 106	18) 26 + 16 = 42	19) 58 + 74 = 132	20) 42 + 39 = 81
21) 72 + 96 = 168	22) 88 + 98 = 186	23) 12 + 77 = 89	24) 85 + 68 = 153	25) 91 + 90 = 181
26) 48 + 17 = 65	27) 44 + 78 = 122	28) 70 + 14 = 84	29) 12 + 83 = 95	30) 29 + 39 = 68
31) 49 + 29 = 78	32) 75 + 46 = 121	33) 40 + 71 = 111	34) 45 + 47 = 92	35) 84 + 82 = 166

ADDITION Page 2

1) 14 + 11 = 25	2) 63 + 11 = 74	3) 66 + 81 = 147	4) 55 + 15 = 70	5) 27 + 14 = 41
6) 21 + 74 = 95	7) 68 + 25 = 93	8) 25 + 50 = 75	9) 29 + 16 = 45	10) 73 + 78 = 151
11) 94 + 88 = 182	12) 12 + 90 = 102	13) 35 + 83 = 118	14) 80 + 87 = 167	15) 48 + 70 = 118
16) 85 + 53 = 138	17) 45 + 86 = 131	18) 27 + 23 = 50	19) 74 + 40 = 114	20) 58 + 69 = 127
21) 38 + 98 = 136	22) 78 + 77 = 155	23) 62 + 37 = 99	24) 53 + 89 = 142	25) 22 + 42 = 64
26) 20 + 39 = 59	27) 35 + 26 = 61	28) 98 + 72 = 170	29) 76 + 82 = 158	30) 11 + 66 = 77
31) 66 + 56 = 122	32) 28 + 67 = 95	33) 39 + 82 = 121	34) 97 + 12 = 109	35) 81 + 15 = 96

ADDITION Page 3

1) 75 + 19 = 94	2) 99 + 65 = 164	3) 94 + 61 = 155	4) 29 + 61 = 90	5) 57 + 32 = 89
6) 19 + 98 = 117	7) 74 + 33 = 107	8) 34 + 31 = 65	9) 96 + 87 = 183	10) 85 + 66 = 151
11) 51 + 15 = 66	12) 39 + 33 = 72	13) 78 + 33 = 111	14) 99 + 19 = 118	15) 51 + 24 = 75
16) 93 + 50 = 143	17) 49 + 53 = 102	18) 67 + 96 = 163	19) 60 + 45 = 105	20) 29 + 44 = 73
21) 29 + 39 = 68	22) 81 + 69 = 150	23) 93 + 82 = 175	24) 16 + 87 = 103	25) 24 + 58 = 82
26) 41 + 63 = 104	27) 32 + 51 = 83	28) 58 + 71 = 129	29) 78 + 76 = 154	30) 86 + 71 = 157
31) 61 + 20 = 81	32) 21 + 43 = 64	33) 37 + 68 = 105	34) 59 + 30 = 89	35) 65 + 37 = 102

ADDITION Page 4

1) 79 + 94 = 173	2) 79 + 13 = 92	3) 47 + 48 = 95	4) 19 + 47 = 66	5) 90 + 55 = 145
6) 49 + 24 = 73	7) 28 + 60 = 88	8) 96 + 50 = 146	9) 60 + 36 = 96	10) 69 + 19 = 88
11) 86 + 13 = 99	12) 21 + 69 = 90	13) 39 + 93 = 132	14) 84 + 38 = 122	15) 15 + 72 = 87
16) 41 + 89 = 130	17) 19 + 16 = 35	18) 13 + 45 = 58	19) 48 + 94 = 142	20) 57 + 57 = 114
21) 67 + 12 = 79	22) 43 + 93 = 136	23) 10 + 87 = 97	24) 44 + 83 = 127	25) 72 + 59 = 131
26) 62 + 77 = 139	27) 11 + 76 = 87	28) 83 + 21 = 104	29) 74 + 53 = 127	30) 63 + 80 = 143
31) 42 + 28 = 70	32) 23 + 59 = 82	33) 16 + 22 = 38	34) 57 + 61 = 118	35) 12 + 65 = 77

ADDITION — Page 5

1) 142 + 90 = 232	2) 149 + 143 = 292	3) 35 + 50 = 85	4) 139 + 62 = 201	5) 23 + 61 = 84
6) 185 + 22 = 207	7) 54 + 170 = 224	8) 109 + 60 = 169	9) 71 + 185 = 256	10) 46 + 189 = 235
11) 22 + 193 = 215	12) 197 + 35 = 232	13) 71 + 152 = 223	14) 166 + 172 = 338	15) 98 + 60 = 158
16) 150 + 175 = 325	17) 135 + 60 = 195	18) 68 + 70 = 138	19) 141 + 32 = 173	20) 23 + 88 = 111
21) 34 + 107 = 141	22) 26 + 30 = 56	23) 114 + 88 = 202	24) 158 + 48 = 206	25) 165 + 146 = 311
26) 44 + 152 = 196	27) 157 + 127 = 284	28) 198 + 150 = 348	29) 93 + 124 = 217	30) 83 + 84 = 167
31) 84 + 167 = 251	32) 164 + 163 = 327	33) 111 + 20 = 131	34) 87 + 194 = 281	35) 112 + 49 = 161

ADDITION — Page 6

1) 195 + 180 = 375	2) 199 + 162 = 361	3) 93 + 173 = 266	4) 117 + 13 = 130	5) 100 + 90 = 190
6) 156 + 160 = 316	7) 22 + 147 = 169	8) 113 + 193 = 306	9) 80 + 142 = 222	10) 194 + 72 = 266
11) 102 + 98 = 200	12) 42 + 154 = 196	13) 109 + 172 = 281	14) 124 + 67 = 191	15) 27 + 183 = 210
16) 40 + 193 = 233	17) 92 + 124 = 216	18) 31 + 29 = 60	19) 28 + 55 = 83	20) 169 + 145 = 314
21) 70 + 24 = 94	22) 60 + 60 = 120	23) 57 + 106 = 163	24) 130 + 116 = 246	25) 17 + 111 = 128
26) 130 + 31 = 161	27) 141 + 64 = 205	28) 136 + 107 = 243	29) 147 + 55 = 202	30) 64 + 120 = 184
31) 186 + 117 = 303	32) 60 + 163 = 223	33) 79 + 98 = 177	34) 126 + 94 = 220	35) 175 + 35 = 210

ADDITION — Page 7

1) 151 + 215 = 366	2) 279 + 132 = 411	3) 252 + 266 = 518	4) 305 + 200 = 505	5) 184 + 139 = 323
6) 413 + 283 = 696	7) 383 + 139 = 522	8) 278 + 295 = 573	9) 390 + 273 = 663	10) 262 + 171 = 433
11) 253 + 162 = 415	12) 482 + 161 = 643	13) 237 + 265 = 502	14) 402 + 299 = 701	15) 349 + 254 = 603
16) 500 + 216 = 716	17) 215 + 204 = 419	18) 382 + 203 = 585	19) 355 + 275 = 630	20) 242 + 266 = 508
21) 138 + 140 = 278	22) 444 + 138 = 582	23) 310 + 149 = 459	24) 430 + 250 = 680	25) 411 + 283 = 694
26) 421 + 213 = 634	27) 192 + 107 = 299	28) 180 + 264 = 444	29) 348 + 128 = 476	30) 426 + 211 = 637
31) 131 + 203 = 334	32) 452 + 188 = 640	33) 178 + 140 = 318	34) 351 + 137 = 488	35) 295 + 286 = 581

ADDITION — Page 8

1) 482 + 111 = 593	2) 363 + 255 = 618	3) 106 + 174 = 280	4) 443 + 173 = 616	5) 387 + 210 = 597
6) 130 + 159 = 289	7) 169 + 195 = 364	8) 404 + 271 = 675	9) 251 + 273 = 524	10) 339 + 106 = 445
11) 436 + 258 = 694	12) 383 + 263 = 646	13) 372 + 273 = 645	14) 112 + 253 = 365	15) 283 + 197 = 480
16) 244 + 248 = 492	17) 249 + 126 = 375	18) 105 + 176 = 281	19) 377 + 195 = 572	20) 223 + 204 = 427
21) 115 + 127 = 242	22) 238 + 150 = 388	23) 373 + 157 = 530	24) 159 + 289 = 448	25) 195 + 194 = 389
26) 196 + 255 = 451	27) 367 + 263 = 630	28) 395 + 133 = 528	29) 239 + 268 = 507	30) 138 + 130 = 268
31) 236 + 119 = 355	32) 462 + 147 = 609	33) 176 + 167 = 343	34) 469 + 108 = 577	35) 284 + 177 = 461

ADDITION — Page 9

1) 235 + 144 = 379	2) 335 + 264 = 599	3) 112 + 245 = 357	4) 457 + 241 = 698	5) 380 + 187 = 567
6) 202 + 214 = 416	7) 467 + 298 = 765	8) 342 + 153 = 495	9) 246 + 230 = 476	10) 346 + 171 = 517
11) 225 + 205 = 430	12) 259 + 280 = 539	13) 120 + 196 = 316	14) 233 + 179 = 412	15) 416 + 106 = 522
16) 344 + 220 = 564	17) 368 + 236 = 604	18) 463 + 173 = 636	19) 293 + 247 = 540	20) 244 + 124 = 368
21) 155 + 129 = 284	22) 371 + 245 = 616	23) 470 + 175 = 645	24) 224 + 125 = 349	25) 282 + 116 = 398
26) 499 + 191 = 690	27) 448 + 200 = 648	28) 499 + 263 = 762	29) 428 + 238 = 666	30) 408 + 210 = 618
31) 257 + 199 = 456	32) 467 + 234 = 701	33) 429 + 232 = 661	34) 459 + 148 = 607	35) 133 + 148 = 281

ADDITION — Page 10

1) 247 + 133 = 380	2) 250 + 157 = 407	3) 431 + 294 = 725	4) 159 + 243 = 402	5) 138 + 175 = 313
6) 173 + 210 = 383	7) 183 + 227 = 410	8) 367 + 286 = 653	9) 392 + 211 = 603	10) 496 + 238 = 734
11) 496 + 233 = 729	12) 291 + 253 = 544	13) 330 + 288 = 618	14) 434 + 254 = 688	15) 383 + 140 = 523
16) 141 + 134 = 275	17) 390 + 183 = 573	18) 367 + 298 = 665	19) 241 + 277 = 518	20) 442 + 125 = 567
21) 260 + 115 = 375	22) 448 + 187 = 635	23) 319 + 118 = 437	24) 107 + 175 = 282	25) 194 + 167 = 361
26) 399 + 203 = 602	27) 120 + 205 = 325	28) 222 + 240 = 462	29) 375 + 222 = 597	30) 385 + 277 = 662
31) 153 + 175 = 328	32) 103 + 279 = 382	33) 210 + 103 = 313	34) 431 + 259 = 690	35) 486 + 236 = 722

ADDITION — Page 11

1) 27 oranges are in the basket. Two more oranges are put in the basket. How many oranges are in the basket now?

   29

2) David has 29 pears and Billy has 11 pears. How many pears do David and Billy have together?

   40

3) Some apples were in the basket. 46 more apples were added to the basket. Now there are 72 apples. How many apples were in the basket before more apples were added?

   26

4) Five red plums and 34 green plums are in the basket. How many plums are in the basket?

   39

5) 62 mangoes were in the basket. 28 are red and the rest are green. How many mangoes are green?

   34

6) 38 bananas were in the basket. More bananas were added to the basket. Now there are 56 bananas. How many bananas were added to the basket?

   18

7) Jennifer has 14 more peaches than Marin. Marin has 39 peaches. How many peaches does Jennifer have?

   53

8) Some plums were in the basket. 39 more plums were added to the basket. Now there are 76 plums. How many plums were in the basket before more plums were added?

   37

ADDITION — Page 12

1) 42 red plums and 49 green plums are in the basket. How many plums are in the basket?

   91

2) 67 mangoes were in the basket. 29 are red and the rest are green. How many mangoes are green?

   38

3) 46 apples were in the basket. More apples were added to the basket. Now there are 48 apples. How many apples were added to the basket?

   2

4) Marcie has 27 more bananas than Ellen. Ellen has 35 bananas. How many bananas does Marcie have?

   62

5) Some pears were in the basket. 38 more pears were added to the basket. Now there are 82 pears. How many pears were in the basket before more pears were added?

   44

6) Billy has 38 peaches and Donald has 30 peaches. How many peaches do Billy and Donald have together?

   68

7) 18 oranges are in the basket. Nine more oranges are put in the basket. How many oranges are in the basket now?

   27

8) 70 plums were in the basket. 20 are red and the rest are green. How many plums are green?

   50

ADDITION — Page 13

1) Donald has 23 peaches and Allan has 46 peaches. How many peaches do Donald and Allan have together?

   69

2) 49 apples are in the basket. 42 more apples are put in the basket. How many apples are in the basket now?

   91

3) 18 bananas were in the basket. More bananas were added to the basket. Now there are 22 bananas. How many bananas were added to the basket?

   4

4) Some mangoes were in the basket. 46 more mangoes were added to the basket. Now there are 79 mangoes. How many mangoes were in the basket before more mangoes were added?

   33

5) Ellen has 32 more oranges than Sharon. Sharon has 23 oranges. How many oranges does Ellen have?

   55

6) Nine red pears and 23 green pears are in the basket. How many pears are in the basket?

   32

7) 18 plums were in the basket. 11 are red and the rest are green. How many plums are green?

   7

8) 41 red plums and 34 green plums are in the basket. How many plums are in the basket?

   75

ADDITION — Page 14

1) 91 mangoes were in the basket. 41 are red and the rest are green. How many mangoes are green?

   50

2) Marin has 24 more peaches than Jackie. Jackie has nine peaches. How many peaches does Marin have?

   33

3) Seven pears were in the basket. More pears were added to the basket. Now there are 40 pears. How many pears were added to the basket?

   33

4) 16 oranges are in the basket. 27 more oranges are put in the basket. How many oranges are in the basket now?

   43

5) Adam has 25 plums and Billy has 44 plums. How many plums do Adam and Billy have together?

   69

6) 23 red apples and three green apples are in the basket. How many apples are in the basket?

   26

7) Some bananas were in the basket. 39 more bananas were added to the basket. Now there are 65 bananas. How many bananas were in the basket before more bananas were added?

   26

8) Jake has 17 mangoes and Paul has 21 mangoes. How many mangoes do Jake and Paul have together?

   38

SUBTRACTION — Page 15

1) 39 − 20 = 19	2) 35 − 31 = 4	3) 12 − 11 = 1	4) 43 − 29 = 14	5) 10 − 10 = 0
6) 39 − 13 = 26	7) 39 − 35 = 4	8) 22 − 18 = 4	9) 31 − 29 = 2	10) 29 − 13 = 16
11) 39 − 15 = 24	12) 25 − 12 = 13	13) 28 − 25 = 3	14) 38 − 26 = 12	15) 44 − 32 = 12
16) 26 − 18 = 8	17) 17 − 15 = 2	18) 36 − 12 = 24	19) 42 − 36 = 6	20) 36 − 18 = 18
21) 41 − 24 = 17	22) 21 − 10 = 11	23) 34 − 24 = 10	24) 35 − 14 = 21	25) 42 − 37 = 5
26) 15 − 15 = 0	27) 37 − 14 = 23	28) 20 − 16 = 4	29) 48 − 30 = 18	30) 25 − 13 = 12
31) 11 − 10 = 1	32) 40 − 21 = 19	33) 39 − 17 = 22	34) 47 − 47 = 0	35) 19 − 12 = 7

SUBTRACTION — Page 16

1) 26 − 21 = 5	2) 29 − 24 = 5	3) 48 − 35 = 13	4) 33 − 10 = 23	5) 30 − 10 = 20
6) 30 − 20 = 10	7) 41 − 36 = 5	8) 23 − 16 = 7	9) 46 − 28 = 18	10) 16 − 10 = 6
11) 34 − 17 = 17	12) 45 − 42 = 3	13) 46 − 31 = 15	14) 49 − 13 = 36	15) 16 − 11 = 5
16) 34 − 24 = 10	17) 17 − 10 = 7	18) 15 − 13 = 2	19) 23 − 14 = 9	20) 31 − 19 = 12
21) 32 − 16 = 16	22) 47 − 25 = 22	23) 40 − 40 = 0	24) 47 − 13 = 34	25) 48 − 45 = 3
26) 39 − 13 = 26	27) 43 − 27 = 16	28) 24 − 12 = 12	29) 43 − 29 = 14	30) 43 − 23 = 20
31) 38 − 24 = 14	32) 44 − 19 = 25	33) 27 − 26 = 1	34) 23 − 20 = 3	35) 39 − 28 = 11

SUBTRACTION — Page 17

1)	2)	3)	4)	5)
70	39	26	73	28
− 42	− 14	− 25	− 41	− 14
28	25	1	32	14
6)	**7)**	**8)**	**9)**	**10)**
87	45	10	73	60
− 50	− 20	− 10	− 16	− 19
37	25	0	57	41
11)	**12)**	**13)**	**14)**	**15)**
95	34	99	58	24
− 50	− 23	− 84	− 52	− 22
45	11	15	6	2
16)	**17)**	**18)**	**19)**	**20)**
51	34	81	67	19
− 37	− 27	− 56	− 58	− 19
14	7	25	9	0
21)	**22)**	**23)**	**24)**	**25)**
34	80	98	33	35
− 10	− 15	− 18	− 31	− 27
24	65	80	2	8
26)	**27)**	**28)**	**29)**	**30)**
76	60	50	33	23
− 32	− 20	− 18	− 28	− 15
44	40	32	5	8
31)	**32)**	**33)**	**34)**	**35)**
48	40	76	32	53
− 42	− 19	− 24	− 25	− 25
6	21	52	7	28

SUBTRACTION — Page 18

1)	2)	3)	4)	5)
87	94	56	15	31
− 37	− 34	− 33	− 11	− 27
50	60	23	4	4
6)	**7)**	**8)**	**9)**	**10)**
82	86	64	26	89
− 28	− 22	− 16	− 17	− 85
54	64	48	9	4
11)	**12)**	**13)**	**14)**	**15)**
78	31	53	29	81
− 22	− 12	− 21	− 26	− 15
56	19	32	3	66
16)	**17)**	**18)**	**19)**	**20)**
10	68	35	23	76
− 10	− 36	− 30	− 20	− 16
0	32	5	3	60
21)	**22)**	**23)**	**24)**	**25)**
41	51	73	83	75
− 31	− 49	− 21	− 33	− 53
10	2	52	50	22
26)	**27)**	**28)**	**29)**	**30)**
68	59	35	20	82
− 17	− 58	− 27	− 18	− 29
51	1	8	2	53
31)	**32)**	**33)**	**34)**	**35)**
36	45	80	63	89
− 15	− 16	− 54	− 32	− 75
21	29	26	31	14

SUBTRACTION — Page 19

1)	2)	3)	4)	5)
33	78	27	84	55
− 20	− 35	− 27	− 44	− 47
13	43	0	40	8
6)	**7)**	**8)**	**9)**	**10)**
44	33	91	92	37
− 12	− 27	− 11	− 30	− 14
32	6	80	62	23
11)	**12)**	**13)**	**14)**	**15)**
69	31	51	46	37
− 35	− 31	− 42	− 14	− 29
34	0	9	32	8
16)	**17)**	**18)**	**19)**	**20)**
12	31	25	87	96
− 11	− 20	− 17	− 34	− 61
1	11	8	53	35
21)	**22)**	**23)**	**24)**	**25)**
85	51	47	13	34
− 74	− 36	− 14	− 12	− 12
11	15	33	1	22
26)	**27)**	**28)**	**29)**	**30)**
69	20	12	26	91
− 19	− 18	− 12	− 22	− 43
50	2	0	4	48
31)	**32)**	**33)**	**34)**	**35)**
62	23	82	83	61
− 48	− 20	− 75	− 68	− 33
14	3	7	15	28

SUBTRACTION — Page 20

1)	2)	3)	4)	5)
198	111	175	178	171
− 129	− 12	− 147	− 150	− 122
69	99	28	28	49
6)	**7)**	**8)**	**9)**	**10)**
177	133	140	119	166
− 39	− 34	− 76	− 67	− 69
138	99	64	52	97
11)	**12)**	**13)**	**14)**	**15)**
166	190	198	126	101
− 66	− 68	− 61	− 77	− 76
100	122	137	49	25
16)	**17)**	**18)**	**19)**	**20)**
161	155	180	184	189
− 93	− 61	− 68	− 76	− 154
68	94	112	108	35
21)	**22)**	**23)**	**24)**	**25)**
164	185	124	140	187
− 95	− 172	− 88	− 35	− 11
69	13	36	105	176
26)	**27)**	**28)**	**29)**	**30)**
123	123	184	173	103
− 67	− 109	− 117	− 137	− 55
56	14	67	36	48
31)	**32)**	**33)**	**34)**	**35)**
157	183	109	163	139
− 39	− 180	− 32	− 160	− 17
118	3	77	3	122

SUBTRACTION Page 21

Problem	Number	Minus	Answer
1)	142	– 125	17
2)	137	– 27	110
3)	173	– 27	146
4)	175	– 125	50
5)	173	– 110	63
6)	150	– 78	72
7)	183	– 93	90
8)	199	– 181	18
9)	195	– 96	99
10)	120	– 59	61
11)	162	– 80	82
12)	123	– 80	43
13)	130	– 56	74
14)	164	– 33	131
15)	107	– 33	74
16)	145	– 46	99
17)	190	– 130	60
18)	108	– 32	76
19)	195	– 110	85
20)	179	– 24	155
21)	186	– 72	114
22)	161	– 130	31
23)	173	– 165	8
24)	189	– 79	110
25)	184	– 29	155
26)	111	– 106	5
27)	160	– 66	94
28)	172	– 165	7
29)	165	– 127	38
30)	183	– 69	114
31)	146	– 92	54
32)	163	– 86	77
33)	116	– 35	81
34)	159	– 35	124
35)	140	– 110	30

SUBTRACTION Page 22

Problem	Number	Minus	Answer
1)	169	– 53	116
2)	148	– 119	29
3)	146	– 24	122
4)	122	– 107	15
5)	147	– 126	21
6)	121	– 43	78
7)	168	– 68	100
8)	166	– 98	68
9)	107	– 31	76
10)	127	– 32	95
11)	176	– 126	50
12)	124	– 67	57
13)	132	– 76	56
14)	173	– 139	34
15)	139	– 139	0
16)	178	– 54	124
17)	123	– 101	22
18)	144	– 40	104
19)	119	– 90	29
20)	149	– 13	136
21)	136	– 51	85
22)	124	– 79	45
23)	141	– 136	5
24)	104	– 81	23
25)	197	– 184	13
26)	129	– 69	60
27)	172	– 154	18
28)	139	– 138	1
29)	104	– 102	2
30)	147	– 135	12
31)	137	– 64	73
32)	148	– 76	72
33)	107	– 102	5
34)	107	– 94	13
35)	193	– 90	103

SUBTRACTION Page 23

Problem	Number	Minus	Answer
1)	243	– 151	92
2)	253	– 170	83
3)	214	– 198	16
4)	225	– 115	110
5)	215	– 124	91
6)	130	– 124	6
7)	248	– 67	181
8)	199	– 88	111
9)	166	– 109	57
10)	128	– 55	73
11)	129	– 47	82
12)	274	– 107	167
13)	293	– 161	132
14)	170	– 158	12
15)	160	– 96	64
16)	113	– 82	31
17)	277	– 162	115
18)	221	– 172	49
19)	111	– 25	86
20)	284	– 177	107
21)	154	– 101	53
22)	211	– 107	104
23)	283	– 113	170
24)	159	– 47	112
25)	213	– 127	86
26)	234	– 23	211
27)	270	– 26	244
28)	219	– 86	133
29)	274	– 58	216
30)	275	– 157	118
31)	228	– 142	86
32)	150	– 11	139
33)	207	– 96	111
34)	113	– 107	6
35)	239	– 132	107

SUBTRACTION Page 24

Problem	Number	Minus	Answer
1)	198	– 76	122
2)	217	– 194	23
3)	116	– 78	38
4)	202	– 134	68
5)	190	– 59	131
6)	207	– 74	133
7)	214	– 30	184
8)	208	– 124	84
9)	153	– 115	38
10)	198	– 134	64
11)	210	– 182	28
12)	214	– 15	199
13)	209	– 35	174
14)	142	– 61	81
15)	127	– 121	6
16)	130	– 77	53
17)	187	– 84	103
18)	182	– 121	61
19)	153	– 129	24
20)	125	– 75	50
21)	194	– 38	156
22)	265	– 175	90
23)	241	– 55	186
24)	262	– 95	167
25)	188	– 163	25
26)	142	– 92	50
27)	112	– 37	75
28)	242	– 122	120
29)	256	– 117	139
30)	185	– 176	9
31)	187	– 32	155
32)	203	– 105	98
33)	125	– 123	2
34)	280	– 38	242
35)	239	– 180	59

SUBTRACTION Page 25

1) 44 apples are in the basket. 43 apples are taken out of the basket. How many apples are in the basket now?

1

2) 69 peaches were in the basket. Some of the peaches were removed from the basket. Now there are 19 peaches. How many peaches were removed from the basket?

50

3) 72 plums are in the basket. 45 are red and the rest are green. How many plums are green?

27

4) Paul has 13 bananas. Brian has 19 bananas. How many more bananas does Brian have than Paul?

6

5) Michele has 50 fewer oranges than Marin. Marin has 66 oranges. How many oranges does Michele have?

16

6) Some pears were in the basket. 27 pears were taken from the basket. Now there are zero pears. How many pears were in the basket before some of the pears were taken?

27

7) Paul has 41 mangoes. Brian has 43 mangoes. How many more mangoes does Brian have than Paul?

2

8) 13 plums were in the basket. Some of the plums were removed from the basket. Now there is one plum. How many plums were removed from the basket?

12

SUBTRACTION Page 26

1) Some apples were in the basket. 26 apples were taken from the basket. Now there are six apples. How many apples were in the basket before some of the apples were taken?

32

2) 11 peaches are in the basket. 11 are red and the rest are green. How many peaches are green?

0

3) Allan has 29 mangoes. Steven has 34 mangoes. How many more mangoes does Steven have than Allan?

5

4) 94 plums were in the basket. Some of the plums were removed from the basket. Now there are 52 plums. How many plums were removed from the basket?

42

5) Amy has 65 fewer oranges than Janet. Janet has 92 oranges. How many oranges does Amy have?

27

6) 20 pears are in the basket. 10 pears are taken out of the basket. How many pears are in the basket now?

10

7) Marcie has 22 fewer bananas than Marin. Marin has 62 bananas. How many bananas does Marcie have?

40

8) Brian has 38 plums. Billy has 54 plums. How many more plums does Billy have than Brian?

16

SUBTRACTION Page 27

1) Jackie has 39 fewer plums than Marin. Marin has 55 plums. How many plums does Jackie have?

16

2) Steven has 32 bananas. Jake has 57 bananas. How many more bananas does Jake have than Steven?

25

3) 51 peaches were in the basket. Some of the peaches were removed from the basket. Now there are 25 peaches. How many peaches were removed from the basket?

26

4) Some apples were in the basket. 35 apples were taken from the basket. Now there are six apples. How many apples were in the basket before some of the apples were taken?

41

5) 32 pears are in the basket. 31 are red and the rest are green. How many pears are green?

1

6) 26 mangoes are in the basket. 25 mangoes are taken out of the basket. How many mangoes are in the basket now?

1

7) Allan has 43 oranges. Paul has 62 oranges. How many more oranges does Paul have than Allan?

19

8) Some pears were in the basket. 36 pears were taken from the basket. Now there are six pears. How many pears were in the basket before some of the pears were taken?

42

SUBTRACTION Page 28

1) 61 apples were in the basket. Some of the apples were removed from the basket. Now there are 50 apples. How many apples were removed from the basket?

11

2) 86 oranges are in the basket. 22 are red and the rest are green. How many oranges are green?

64

3) Amy has 50 fewer pears than Jackie. Jackie has 88 pears. How many pears does Amy have?

38

4) Some peaches were in the basket. 11 peaches were taken from the basket. Now there are three peaches. How many peaches were in the basket before some of the peaches were taken?

14

5) 58 mangoes are in the basket. 24 mangoes are taken out of the basket. How many mangoes are in the basket now?

34

6) Donald has 24 bananas. Jake has 29 bananas. How many more bananas does Jake have than Donald?

5

7) Marin has seven fewer plums than Marcie. Marcie has 46 plums. How many plums does Marin have?

39

8) Some bananas were in the basket. 25 bananas were taken from the basket. Now there are 41 bananas. How many bananas were in the basket before some of the bananas were taken?

66

MULTIPLICATION — Page 29

1) 4 × 9 = 36	2) 5 × 7 = 35	3) 7 × 3 = 21	4) 7 × 6 = 42	5) 8 × 4 = 32
6) 2 × 3 = 6	7) 7 × 8 = 56	8) 3 × 9 = 27	9) 4 × 6 = 24	10) 5 × 3 = 15
11) 5 × 6 = 30	12) 5 × 8 = 40	13) 6 × 5 = 30	14) 3 × 4 = 12	15) 8 × 10 = 80
16) 8 × 6 = 48	17) 8 × 7 = 56	18) 10 × 4 = 40	19) 3 × 8 = 24	20) 5 × 10 = 50
21) 8 × 3 = 24	22) 3 × 3 = 9	23) 4 × 2 = 8	24) 8 × 9 = 72	25) 9 × 7 = 63
26) 10 × 5 = 50	27) 8 × 8 = 64	28) 7 × 5 = 35	29) 7 × 4 = 28	30) 6 × 3 = 18
31) 10 × 7 = 70	32) 6 × 7 = 42	33) 9 × 3 = 27	34) 7 × 10 = 70	35) 3 × 6 = 18

MULTIPLICATION — Page 30

1) 6 × 7 = 42	2) 7 × 3 = 21	3) 5 × 2 = 10	4) 5 × 4 = 20	5) 3 × 7 = 21
6) 7 × 4 = 28	7) 10 × 6 = 60	8) 6 × 4 = 24	9) 4 × 6 = 24	10) 9 × 9 = 81
11) 8 × 7 = 56	12) 4 × 10 = 40	13) 7 × 10 = 70	14) 9 × 8 = 72	15) 6 × 9 = 54
16) 8 × 6 = 48	17) 3 × 10 = 30	18) 5 × 8 = 40	19) 2 × 9 = 18	20) 6 × 6 = 36
21) 2 × 6 = 12	22) 6 × 8 = 48	23) 4 × 5 = 20	24) 9 × 6 = 54	25) 2 × 5 = 10
26) 8 × 9 = 72	27) 10 × 8 = 80	28) 8 × 8 = 64	29) 8 × 4 = 32	30) 7 × 6 = 42
31) 9 × 2 = 18	32) 10 × 3 = 30	33) 7 × 7 = 49	34) 7 × 9 = 63	35) 9 × 10 = 90

MULTIPLICATION — Page 31

1) 9 × 7 = 63	2) 47 × 5 = 235	3) 20 × 4 = 80	4) 45 × 6 = 270	5) 34 × 7 = 238
6) 20 × 9 = 180	7) 47 × 4 = 188	8) 9 × 5 = 45	9) 13 × 3 = 39	10) 19 × 9 = 171
11) 33 × 6 = 198	12) 19 × 5 = 95	13) 27 × 4 = 108	14) 39 × 9 = 351	15) 28 × 3 = 84
16) 21 × 7 = 147	17) 34 × 3 = 102	18) 20 × 3 = 60	19) 5 × 6 = 30	20) 14 × 6 = 84
21) 4 × 8 = 32	22) 45 × 10 = 450	23) 32 × 10 = 320	24) 34 × 5 = 170	25) 48 × 4 = 192
26) 15 × 8 = 120	27) 23 × 8 = 184	28) 17 × 6 = 102	29) 43 × 3 = 129	30) 23 × 3 = 69
31) 46 × 7 = 322	32) 12 × 8 = 96	33) 10 × 9 = 90	34) 44 × 3 = 132	35) 16 × 7 = 112

MULTIPLICATION — Page 32

1) 38 × 3 = 114	2) 13 × 4 = 52	3) 38 × 10 = 380	4) 16 × 6 = 96	5) 33 × 9 = 297
6) 40 × 6 = 240	7) 50 × 5 = 250	8) 16 × 4 = 64	9) 12 × 4 = 48	10) 33 × 3 = 99
11) 22 × 7 = 154	12) 48 × 10 = 480	13) 18 × 8 = 144	14) 44 × 7 = 308	15) 29 × 2 = 58
16) 42 × 5 = 210	17) 33 × 7 = 231	18) 21 × 5 = 105	19) 2 × 7 = 14	20) 21 × 3 = 63
21) 38 × 4 = 152	22) 37 × 2 = 74	23) 31 × 9 = 279	24) 26 × 6 = 156	25) 46 × 4 = 184
26) 20 × 4 = 80	27) 13 × 7 = 91	28) 38 × 7 = 266	29) 17 × 8 = 136	30) 33 × 6 = 198
31) 31 × 10 = 310	32) 19 × 8 = 152	33) 32 × 7 = 224	34) 35 × 7 = 245	35) 43 × 8 = 344

MULTIPLICATION — Page 33

1) 49 × 6 = 294	2) 32 × 3 = 96	3) 75 × 7 = 525	4) 93 × 9 = 837	5) 34 × 10 = 340
6) 52 × 6 = 312	7) 77 × 3 = 231	8) 20 × 3 = 60	9) 79 × 4 = 316	10) 71 × 6 = 426
11) 91 × 3 = 273	12) 90 × 7 = 630	13) 78 × 8 = 624	14) 83 × 10 = 830	15) 71 × 8 = 568
16) 87 × 6 = 522	17) 27 × 9 = 243	18) 13 × 3 = 39	19) 88 × 5 = 440	20) 33 × 8 = 264
21) 28 × 4 = 112	22) 49 × 8 = 392	23) 63 × 3 = 189	24) 24 × 8 = 192	25) 73 × 7 = 511
26) 79 × 2 = 158	27) 21 × 4 = 84	28) 77 × 4 = 308	29) 65 × 7 = 455	30) 46 × 2 = 92
31) 79 × 10 = 790	32) 25 × 6 = 150	33) 16 × 10 = 160	34) 36 × 9 = 324	35) 22 × 2 = 44

MULTIPLICATION — Page 34

1) 40 × 2 = 80	2) 29 × 2 = 58	3) 15 × 6 = 90	4) 55 × 7 = 385	5) 76 × 6 = 456
6) 70 × 8 = 560	7) 16 × 7 = 112	8) 32 × 6 = 192	9) 44 × 8 = 352	10) 43 × 6 = 258
11) 97 × 6 = 582	12) 79 × 9 = 711	13) 14 × 7 = 98	14) 84 × 5 = 420	15) 41 × 9 = 369
16) 58 × 8 = 464	17) 65 × 8 = 520	18) 27 × 9 = 243	19) 18 × 4 = 72	20) 97 × 7 = 679
21) 27 × 8 = 216	22) 39 × 6 = 234	23) 98 × 4 = 392	24) 52 × 3 = 156	25) 49 × 7 = 343
26) 28 × 3 = 84	27) 47 × 9 = 423	28) 32 × 4 = 128	29) 96 × 5 = 480	30) 73 × 3 = 219
31) 37 × 9 = 333	32) 29 × 5 = 145	33) 96 × 6 = 576	34) 41 × 6 = 246	35) 64 × 3 = 192

MULTIPLICATION — Page 35

1) 18 × 5 = 90	2) 24 × 3 = 72	3) 182 × 10 = 1820	4) 58 × 6 = 348	5) 51 × 3 = 153
6) 63 × 10 = 630	7) 99 × 6 = 594	8) 188 × 6 = 1128	9) 150 × 3 = 450	10) 82 × 3 = 246
11) 61 × 5 = 305	12) 117 × 9 = 1053	13) 199 × 6 = 1194	14) 70 × 6 = 420	15) 74 × 4 = 296
16) 98 × 7 = 686	17) 153 × 5 = 765	18) 28 × 8 = 224	19) 95 × 8 = 760	20) 200 × 5 = 1000
21) 116 × 9 = 1044	22) 138 × 4 = 552	23) 181 × 5 = 905	24) 193 × 3 = 579	25) 170 × 7 = 1190
26) 17 × 5 = 85	27) 90 × 4 = 360	28) 50 × 7 = 350	29) 157 × 4 = 628	30) 34 × 5 = 170
31) 136 × 8 = 1088	32) 46 × 9 = 414	33) 145 × 7 = 1015	34) 55 × 3 = 165	35) 80 × 3 = 240

MULTIPLICATION — Page 36

1) 12 × 4 = 48	2) 97 × 7 = 679	3) 160 × 10 = 1600	4) 109 × 4 = 436	5) 98 × 8 = 784
6) 102 × 4 = 408	7) 94 × 7 = 658	8) 118 × 5 = 590	9) 138 × 2 = 276	10) 187 × 4 = 748
11) 162 × 6 = 972	12) 96 × 4 = 384	13) 156 × 8 = 1248	14) 191 × 7 = 1337	15) 136 × 7 = 952
16) 159 × 5 = 795	17) 191 × 3 = 573	18) 58 × 9 = 522	19) 157 × 9 = 1413	20) 70 × 6 = 420
21) 97 × 9 = 873	22) 34 × 9 = 306	23) 62 × 7 = 434	24) 51 × 3 = 153	25) 16 × 9 = 144
26) 90 × 8 = 720	27) 169 × 6 = 1014	28) 158 × 7 = 1106	29) 154 × 4 = 616	30) 180 × 2 = 360
31) 199 × 9 = 1791	32) 132 × 9 = 1188	33) 193 × 3 = 579	34) 56 × 4 = 224	35) 86 × 7 = 602

MULTIPLICATION Page 37

1) 168 × 5 = 840
2) 127 × 5 = 635
3) 197 × 9 = 1773
4) 173 × 10 = 1730
5) 155 × 10 = 1550
6) 156 × 6 = 936
7) 132 × 9 = 1188
8) 171 × 9 = 1539
9) 165 × 7 = 1155
10) 157 × 7 = 1099
11) 112 × 2 = 224
12) 114 × 4 = 456
13) 195 × 8 = 1560
14) 127 × 6 = 762
15) 116 × 5 = 580
16) 159 × 7 = 1113
17) 164 × 9 = 1476
18) 148 × 9 = 1332
19) 111 × 3 = 333
20) 140 × 9 = 1260
21) 154 × 4 = 616
22) 169 × 5 = 845
23) 109 × 10 = 1090
24) 159 × 6 = 954
25) 164 × 3 = 492
26) 156 × 5 = 780
27) 184 × 10 = 1840
28) 153 × 7 = 1071
29) 130 × 2 = 260
30) 148 × 4 = 592
31) 159 × 4 = 636
32) 182 × 3 = 546
33) 141 × 10 = 1410
34) 189 × 5 = 945
35) 138 × 9 = 1242

MULTIPLICATION Page 38

1) 109 × 4 = 436
2) 115 × 9 = 1035
3) 170 × 4 = 680
4) 152 × 9 = 1368
5) 150 × 3 = 450
6) 179 × 6 = 1074
7) 182 × 4 = 728
8) 124 × 4 = 496
9) 124 × 2 = 248
10) 152 × 4 = 608
11) 175 × 8 = 1400
12) 106 × 5 = 530
13) 186 × 5 = 930
14) 182 × 9 = 1638
15) 199 × 9 = 1791
16) 196 × 8 = 1568
17) 102 × 7 = 714
18) 177 × 10 = 1770
19) 161 × 8 = 1288
20) 168 × 7 = 1176
21) 104 × 3 = 312
22) 121 × 2 = 242
23) 150 × 9 = 1350
24) 164 × 4 = 656
25) 171 × 10 = 1710
26) 185 × 3 = 555
27) 158 × 6 = 948
28) 160 × 5 = 800
29) 114 × 8 = 912
30) 100 × 5 = 500
31) 144 × 8 = 1152
32) 192 × 7 = 1344
33) 115 × 3 = 345
34) 143 × 3 = 429
35) 153 × 4 = 612

MULTIPLICATION Page 39

1) If there are 5 bananas in each box and there are 5 boxes, how many bananas are there in total?
25

2) Steven can cycle 7 miles per hour. How far can Steven cycle in 9 hours?
63

3) Paul has 19 times more oranges than Janet. Janet has 6 oranges. How many oranges does Paul have?
114

4) David swims 18 laps every day. How many laps will David swim in 3 days?
54

5) Amy's garden has 13 rows of pumpkins. Each row has 5 pumpkins. How many pumpkins does Amy have in all?
65

6) If there are 20 peaches in each box and there are 3 boxes, how many peaches are there in total?
60

7) Audrey has 11 times more mangoes than Sharon. Sharon has 7 mangoes. How many mangoes does Audrey have?
77

8) Michele's garden has 19 rows of pumpkins. Each row has 2 pumpkins. How many pumpkins does Michele have in all?
38

MULTIPLICATION Page 40

1) Janet has 20 times more peaches than David. David has 7 peaches. How many peaches does Janet have?
140

2) If there are 35 oranges in each box and there are 2 boxes, how many oranges are there in total?
70

3) Brian can cycle 3 miles per hour. How far can Brian cycle in 2 hours?
6

4) Amy's garden has 43 rows of pumpkins. Each row has 3 pumpkins. How many pumpkins does Amy have in all?
129

5) Jackie swims 10 laps every day. How many laps will Jackie swim in 5 days?
50

6) If there are 33 bananas in each box and there are 3 boxes, how many bananas are there in total?
99

7) Adam swims 40 laps every day. How many laps will Adam swim in 3 days?
120

8) Sharon's garden has 36 rows of pumpkins. Each row has 9 pumpkins. How many pumpkins does Sharon have in all?
324

MULTIPLICATION Page 41

1) Adam has 43 times more peaches than Brian. Brian has 4 peaches. How many peaches does Adam have?

172

2) Billy can cycle 6 miles per hour. How far can Billy cycle in 3 hours?

18

3) If there are 28 oranges in each box and there are 7 boxes, how many oranges are there in total?

196

4) Steven swims 12 laps every day. How many laps will Steven swim in 6 days?

72

5) Jackie's garden has 39 rows of pumpkins. Each row has 3 pumpkins. How many pumpkins does Jackie have in all?

117

6) Jake swims 13 laps every day. How many laps will Jake swim in 6 days?

78

7) Jake can cycle 17 miles per hour. How far can Jake cycle in 9 hours?

153

8) Amy's garden has 11 rows of pumpkins. Each row has 10 pumpkins. How many pumpkins does Amy have in all?

110

MULTIPLICATION Page 42

1) Ellen's garden has 18 rows of pumpkins. Each row has 2 pumpkins. How many pumpkins does Ellen have in all?

36

2) Jackie has 41 times more apples than Janet. Janet has 9 apples. How many apples does Jackie have?

369

3) If there are 47 plums in each box and there are 9 boxes, how many plums are there in total?

423

4) Adam swims 5 laps every day. How many laps will Adam swim in 5 days?

25

5) Steven can cycle 41 miles per hour. How far can Steven cycle in 10 hours?

410

6) If there are 3 peaches in each box and there are 9 boxes, how many peaches are there in total?

27

7) Steven has 3 times more bananas than Ellen. Ellen has 6 bananas. How many bananas does Steven have?

18

8) Allan can cycle 16 miles per hour. How far can Allan cycle in 5 hours?

80

DIVISION Page 43

1) 25 ÷ 5 = 5	2) 4 ÷ 2 = 2	3) 56 ÷ 8 = 7
4) 6 ÷ 3 = 2	5) 32 ÷ 4 = 8	6) 27 ÷ 9 = 3
7) 48 ÷ 8 = 6	8) 48 ÷ 6 = 8	9) 54 ÷ 9 = 6
10) 40 ÷ 10 = 4	11) 24 ÷ 3 = 8	12) 3 ÷ 3 = 1
13) 27 ÷ 3 = 9	14) 42 ÷ 6 = 7	15) 49 ÷ 7 = 7
16) 40 ÷ 5 = 8	17) 7 ÷ 7 = 1	18) 35 ÷ 7 = 5
19) 18 ÷ 3 = 6	20) 50 ÷ 10 = 5	21) 36 ÷ 9 = 4
22) 72 ÷ 8 = 9	23) 6 ÷ 6 = 1	24) 30 ÷ 10 = 3
25) 20 ÷ 4 = 5	26) 5 ÷ 5 = 1	27) 70 ÷ 10 = 7
28) 16 ÷ 4 = 4	29) 8 ÷ 4 = 2	30) 72 ÷ 9 = 8
31) 10 ÷ 2 = 5	32) 24 ÷ 6 = 4	33) 12 ÷ 4 = 3
34) 21 ÷ 3 = 7	35) 16 ÷ 8 = 2	36) 12 ÷ 6 = 2

DIVISION Page 44

1) 18 ÷ 9 = 2	2) 15 ÷ 3 = 5	3) 12 ÷ 4 = 3
4) 20 ÷ 5 = 4	5) 21 ÷ 7 = 3	6) 4 ÷ 4 = 1
7) 10 ÷ 2 = 5	8) 54 ÷ 6 = 9	9) 45 ÷ 9 = 5
10) 14 ÷ 7 = 2	11) 4 ÷ 2 = 2	12) 32 ÷ 4 = 8
13) 9 ÷ 9 = 1	14) 36 ÷ 6 = 6	15) 36 ÷ 9 = 4
16) 30 ÷ 5 = 6	17) 18 ÷ 2 = 9	18) 35 ÷ 7 = 5
19) 16 ÷ 4 = 4	20) 64 ÷ 8 = 8	21) 48 ÷ 6 = 8
22) 18 ÷ 3 = 6	23) 12 ÷ 3 = 4	24) 12 ÷ 2 = 6
25) 27 ÷ 3 = 9	26) 24 ÷ 3 = 8	27) 28 ÷ 7 = 4
28) 28 ÷ 4 = 7	29) 25 ÷ 5 = 5	30) 81 ÷ 9 = 9
31) 24 ÷ 6 = 4	32) 72 ÷ 9 = 8	33) 56 ÷ 8 = 7
34) 3 ÷ 3 = 1	35) 42 ÷ 6 = 7	36) 60 ÷ 10 = 6

DIVISION Page 45

1) 10 ÷ 2 = 5	2) 28 ÷ 4 = 7	3) 9 ÷ 3 = 3
4) 20 ÷ 4 = 5	5) 12 ÷ 2 = 6	6) 42 ÷ 6 = 7
7) 27 ÷ 9 = 3	8) 28 ÷ 7 = 4	9) 25 ÷ 5 = 5
10) 6 ÷ 6 = 1	11) 18 ÷ 3 = 6	12) 32 ÷ 4 = 8
13) 40 ÷ 8 = 5	14) 35 ÷ 5 = 7	15) 30 ÷ 5 = 6
16) 20 ÷ 10 = 2	17) 64 ÷ 8 = 8	18) 45 ÷ 9 = 5
19) 12 ÷ 6 = 2	20) 63 ÷ 9 = 7	21) 12 ÷ 3 = 4
22) 27 ÷ 3 = 9	23) 24 ÷ 6 = 4	24) 16 ÷ 4 = 4
25) 21 ÷ 3 = 7	26) 54 ÷ 9 = 6	27) 63 ÷ 7 = 9
28) 7 ÷ 7 = 1	29) 72 ÷ 9 = 8	30) 56 ÷ 7 = 8
31) 35 ÷ 7 = 5	32) 21 ÷ 7 = 3	33) 70 ÷ 10 = 7
34) 50 ÷ 10 = 5	35) 4 ÷ 2 = 2	36) 15 ÷ 5 = 3

DIVISION Page 46

1) 16 ÷ 8 = 2	2) 7 ÷ 7 = 1	3) 27 ÷ 9 = 3
4) 14 ÷ 7 = 2	5) 36 ÷ 6 = 6	6) 8 ÷ 8 = 1
7) 72 ÷ 8 = 9	8) 6 ÷ 2 = 3	9) 12 ÷ 3 = 4
10) 35 ÷ 7 = 5	11) 42 ÷ 7 = 6	12) 12 ÷ 6 = 2
13) 45 ÷ 9 = 5	14) 30 ÷ 5 = 6	15) 21 ÷ 3 = 7
16) 10 ÷ 5 = 2	17) 18 ÷ 6 = 3	18) 9 ÷ 3 = 3
19) 32 ÷ 4 = 8	20) 24 ÷ 6 = 4	21) 8 ÷ 4 = 2
22) 2 ÷ 2 = 1	23) 15 ÷ 3 = 5	24) 63 ÷ 7 = 9
25) 12 ÷ 4 = 3	26) 6 ÷ 6 = 1	27) 42 ÷ 6 = 7
28) 56 ÷ 8 = 7	29) 48 ÷ 6 = 8	30) 48 ÷ 8 = 6
31) 18 ÷ 3 = 6	32) 20 ÷ 10 = 2	33) 25 ÷ 5 = 5
34) 40 ÷ 5 = 8	35) 4 ÷ 2 = 2	36) 28 ÷ 7 = 4

DIVISION Page 47

1) 18 ÷ 3 = 6	2) 2 ÷ 2 = 1	3) 6 ÷ 6 = 1
4) 45 ÷ 9 = 5	5) 35 ÷ 5 = 7	6) 42 ÷ 7 = 6
7) 20 ÷ 5 = 4	8) 49 ÷ 7 = 7	9) 21 ÷ 7 = 3
10) 56 ÷ 8 = 7	11) 63 ÷ 7 = 9	12) 16 ÷ 4 = 4
13) 8 ÷ 2 = 4	14) 24 ÷ 4 = 6	15) 36 ÷ 4 = 9
16) 27 ÷ 3 = 9	17) 21 ÷ 3 = 7	18) 42 ÷ 6 = 7
19) 28 ÷ 4 = 7	20) 64 ÷ 8 = 8	21) 36 ÷ 9 = 4
22) 12 ÷ 3 = 4	23) 70 ÷ 10 = 7	24) 35 ÷ 7 = 5
25) 4 ÷ 2 = 2	26) 16 ÷ 8 = 2	27) 14 ÷ 2 = 7
28) 15 ÷ 5 = 3	29) 32 ÷ 4 = 8	30) 12 ÷ 2 = 6
31) 60 ÷ 10 = 6	32) 40 ÷ 8 = 5	33) 12 ÷ 6 = 2
34) 30 ÷ 10 = 3	35) 25 ÷ 5 = 5	36) 48 ÷ 8 = 6

DIVISION Page 48

1) 30 ÷ 6 = 5	2) 8 ÷ 2 = 4	3) 30 ÷ 15 = 2
4) 68 ÷ 17 = 4	5) 60 ÷ 12 = 5	6) 28 ÷ 4 = 7
7) 49 ÷ 7 = 7	8) 34 ÷ 17 = 2	9) 10 ÷ 5 = 2
10) 112 ÷ 14 = 8	11) 30 ÷ 10 = 3	12) 128 ÷ 16 = 8
13) 32 ÷ 4 = 8	14) 4 ÷ 4 = 1	15) 38 ÷ 19 = 2
16) 95 ÷ 19 = 5	17) 48 ÷ 16 = 3	18) 84 ÷ 14 = 6
19) 7 ÷ 7 = 1	20) 8 ÷ 4 = 2	21) 54 ÷ 9 = 6
22) 28 ÷ 14 = 2	23) 21 ÷ 7 = 3	24) 85 ÷ 17 = 5
25) 80 ÷ 20 = 4	26) 20 ÷ 4 = 5	27) 9 ÷ 3 = 3
28) 45 ÷ 5 = 9	29) 45 ÷ 15 = 3	30) 56 ÷ 7 = 8
31) 133 ÷ 19 = 7	32) 40 ÷ 10 = 4	33) 66 ÷ 11 = 6
34) 2 ÷ 2 = 1	35) 6 ÷ 3 = 2	36) 36 ÷ 18 = 2

DIVISION Page 49

1) 60 ÷ 15 = 4 2) 18 ÷ 9 = 2 3) 56 ÷ 14 = 4

4) 72 ÷ 18 = 4 5) 120 ÷ 15 = 8 6) 81 ÷ 9 = 9

7) 35 ÷ 7 = 5 8) 76 ÷ 19 = 4 9) 42 ÷ 6 = 7

10) 72 ÷ 9 = 8 11) 48 ÷ 6 = 8 12) 152 ÷ 19 = 8

13) 45 ÷ 15 = 3 14) 12 ÷ 4 = 3 15) 8 ÷ 4 = 2

16) 21 ÷ 3 = 7 17) 119 ÷ 17 = 7 18) 56 ÷ 7 = 8

19) 40 ÷ 10 = 4 20) 54 ÷ 18 = 3 21) 48 ÷ 8 = 6

22) 32 ÷ 16 = 2 23) 84 ÷ 12 = 7 24) 72 ÷ 12 = 6

25) 14 ÷ 14 = 1 26) 28 ÷ 4 = 7 27) 84 ÷ 14 = 6

28) 28 ÷ 7 = 4 29) 64 ÷ 16 = 4 30) 95 ÷ 19 = 5

31) 42 ÷ 14 = 3 32) 12 ÷ 3 = 4 33) 16 ÷ 8 = 2

34) 48 ÷ 12 = 4 35) 8 ÷ 2 = 4 36) 36 ÷ 12 = 3

DIVISION Page 50

1) 28 ÷ 7 = 4 2) 60 ÷ 15 = 4 3) 85 ÷ 17 = 5

4) 28 ÷ 4 = 7 5) 78 ÷ 13 = 6 6) 81 ÷ 9 = 9

7) 96 ÷ 16 = 6 8) 36 ÷ 18 = 2 9) 77 ÷ 11 = 7

10) 24 ÷ 3 = 8 11) 38 ÷ 19 = 2 12) 36 ÷ 9 = 4

13) 32 ÷ 4 = 8 14) 128 ÷ 16 = 8 15) 22 ÷ 11 = 2

16) 24 ÷ 8 = 3 17) 30 ÷ 15 = 2 18) 42 ÷ 14 = 3

19) 45 ÷ 15 = 3 20) 40 ÷ 10 = 4 21) 91 ÷ 13 = 7

22) 6 ÷ 3 = 2 23) 18 ÷ 6 = 3 24) 45 ÷ 9 = 5

25) 48 ÷ 12 = 4 26) 33 ÷ 11 = 3 27) 70 ÷ 10 = 7

28) 133 ÷ 19 = 7 29) 16 ÷ 4 = 4 30) 65 ÷ 13 = 5

31) 136 ÷ 17 = 8 32) 20 ÷ 10 = 2 33) 14 ÷ 2 = 7

34) 9 ÷ 3 = 3 35) 35 ÷ 5 = 7 36) 44 ÷ 11 = 4

DIVISION Page 51

1) 175 ÷ 25 = 7 2) 161 ÷ 23 = 7 3) 144 ÷ 24 = 6

4) 171 ÷ 19 = 9 5) 84 ÷ 21 = 4 6) 133 ÷ 19 = 7

7) 90 ÷ 15 = 6 8) 48 ÷ 16 = 3 9) 76 ÷ 19 = 4

10) 77 ÷ 11 = 7 11) 52 ÷ 13 = 4 12) 30 ÷ 10 = 3

13) 135 ÷ 15 = 9 14) 154 ÷ 22 = 7 15) 72 ÷ 12 = 6

16) 240 ÷ 30 = 8 17) 65 ÷ 13 = 5 18) 38 ÷ 19 = 2

19) 153 ÷ 17 = 9 20) 56 ÷ 28 = 2 21) 48 ÷ 24 = 2

22) 23 ÷ 23 = 1 23) 68 ÷ 17 = 4 24) 84 ÷ 12 = 7

25) 140 ÷ 28 = 5 26) 98 ÷ 14 = 7 27) 36 ÷ 18 = 2

28) 44 ÷ 11 = 4 29) 96 ÷ 24 = 4 30) 147 ÷ 21 = 7

31) 174 ÷ 29 = 6 32) 216 ÷ 27 = 8 33) 102 ÷ 17 = 6

34) 30 ÷ 15 = 2 35) 145 ÷ 29 = 5 36) 152 ÷ 19 = 8

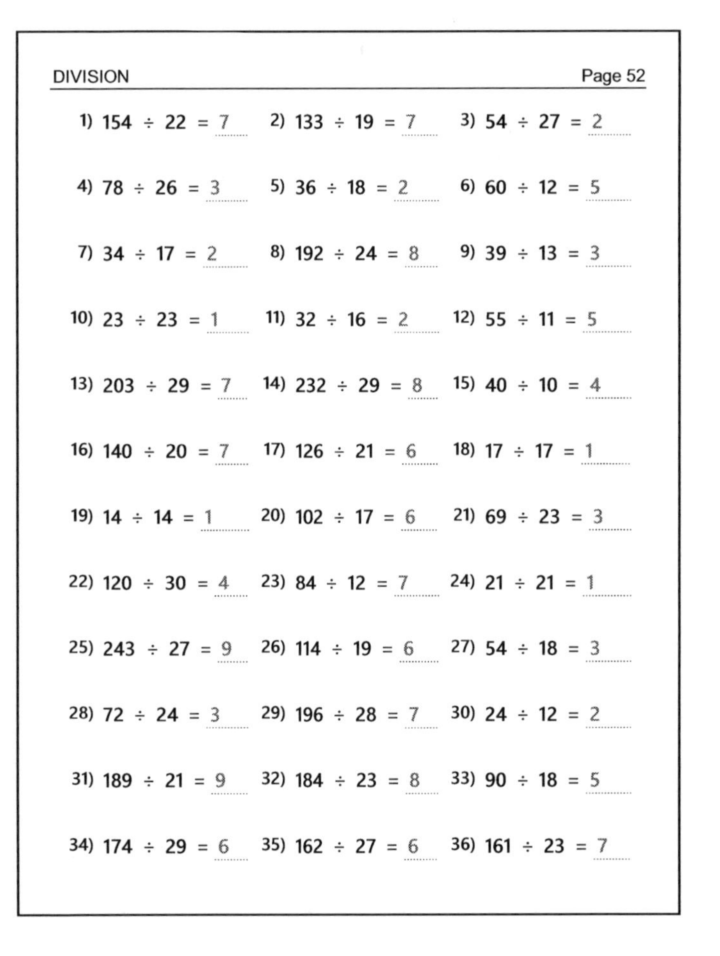

DIVISION Page 52

1) 154 ÷ 22 = 7 2) 133 ÷ 19 = 7 3) 54 ÷ 27 = 2

4) 78 ÷ 26 = 3 5) 36 ÷ 18 = 2 6) 60 ÷ 12 = 5

7) 34 ÷ 17 = 2 8) 192 ÷ 24 = 8 9) 39 ÷ 13 = 3

10) 23 ÷ 23 = 1 11) 32 ÷ 16 = 2 12) 55 ÷ 11 = 5

13) 203 ÷ 29 = 7 14) 232 ÷ 29 = 8 15) 40 ÷ 10 = 4

16) 140 ÷ 20 = 7 17) 126 ÷ 21 = 6 18) 17 ÷ 17 = 1

19) 14 ÷ 14 = 1 20) 102 ÷ 17 = 6 21) 69 ÷ 23 = 3

22) 120 ÷ 30 = 4 23) 84 ÷ 12 = 7 24) 21 ÷ 21 = 1

25) 243 ÷ 27 = 9 26) 114 ÷ 19 = 6 27) 54 ÷ 18 = 3

28) 72 ÷ 24 = 3 29) 196 ÷ 28 = 7 30) 24 ÷ 12 = 2

31) 189 ÷ 21 = 9 32) 184 ÷ 23 = 8 33) 90 ÷ 18 = 5

34) 174 ÷ 29 = 6 35) 162 ÷ 27 = 6 36) 161 ÷ 23 = 7

1) Michele made 160 cookies for a bake sale. She put the cookies in bags, with 20 cookies in each bag. How many bags did she have for the bake sale?

8

2) Sandra ordered nine pizzas. The bill for the pizzas came to $126. What was the cost of each pizza?

14

3) A box of bananas weighs 360 pounds. If one banana weighs 20 pounds, how many bananas are there in the box?

18

4) How many 13 cm pieces of rope can you cut from a rope that is 195 cm long?

15

5) You have 70 plums and want to share them equally with 14 people. How many plums would each person get?

5

6) Brian is reading a book with 170 pages. If Brian wants to read the same number of pages every day, how many pages would Brian have to read each day to finish in 10 days?

17

7) Steven is reading a book with 175 pages. If Steven wants to read the same number of pages every day, how many pages would Steven have to read each day to finish in 25 days?

7

8) How many 26 cm pieces of rope can you cut from a rope that is 234 cm long?

9

1) A box of oranges weighs 105 pounds. If one orange weighs seven pounds, how many oranges are there in the box?

15

2) You have 416 pears and want to share them equally with 26 people. How many pears would each person get?

16

3) How many 12 cm pieces of rope can you cut from a rope that is 216 cm long?

18

4) Billy is reading a book with 560 pages. If Billy wants to read the same number of pages every day, how many pages would Billy have to read each day to finish in 28 days?

20

5) Steven ordered eight pizzas. The bill for the pizzas came to $112. What was the cost of each pizza?

14

6) Jackie made 143 cookies for a bake sale. She put the cookies in bags, with 11 cookies in each bag. How many bags did she have for the bake sale?

13

7) Billy is reading a book with 180 pages. If Billy wants to read the same number of pages every day, how many pages would Billy have to read each day to finish in 20 days?

9

8) Amy ordered 15 pizzas. The bill for the pizzas came to $105. What was the cost of each pizza?

7

1) David is reading a book with 459 pages. If David wants to read the same number of pages every day, how many pages would David have to read each day to finish in 27 days?

17

2) Marin ordered 27 pizzas. The bill for the pizzas came to $540. What was the cost of each pizza?

20

3) How many 17 cm pieces of rope can you cut from a rope that is 204 cm long?

12

4) You have 110 mangoes and want to share them equally with 10 people. How many mangoes would each person get?

11

5) Sharon made 57 cookies for a bake sale. She put the cookies in bags, with 19 cookies in each bag. How many bags did she have for the bake sale?

3

6) A box of peaches weighs 208 pounds. If one peach weighs 13 pounds, how many peaches are there in the box?

16

7) Jackie ordered 17 pizzas. The bill for the pizzas came to $102. What was the cost of each pizza?

6

8) You have 209 plums and want to share them equally with 19 people. How many plums would each person get?

11

1) A box of pears weighs 39 pounds. If one pear weighs three pounds, how many pears are there in the box?

13

2) How many 20 cm pieces of rope can you cut from a rope that is 140 cm long?

7

3) Marin ordered 19 pizzas. The bill for the pizzas came to $228. What was the cost of each pizza?

12

4) Allan is reading a book with 51 pages. If Allan wants to read the same number of pages every day, how many pages would Allan have to read each day to finish in three days?

17

5) You have 90 apples and want to share them equally with 15 people. How many apples would each person get?

6

6) Marcie made 266 cookies for a bake sale. She put the cookies in bags, with 14 cookies in each bag. How many bags did she have for the bake sale?

19

7) A box of mangoes weighs 256 pounds. If one mango weighs 16 pounds, how many mangoes are there in the box?

16

8) Allan is reading a book with eight pages. If Allan wants to read the same number of pages every day, how many pages would Allan have to read each day to finish in four days?

2

Made in the USA
Las Vegas, NV
30 November 2023